Nineteen Ways

of

Looking at
Consciousness

Nineteen Ways

of

Looking at
Consciousness

PATRICK HOUSE

ST. MARTIN'S GRIFFIN
NEW YORK

Published in the United States by St. Martin's Griffin, an imprint of St. Martin's Publishing Group

NINETEEN WAYS OF LOOKING AT CONSCIOUSNESS. Copyright © 2022 by Patrick House. All rights reserved. Printed in the United States of America. For information, address St. Martin's Publishing Group, 120 Broadway, New York, NY 10271.

www.stmartins.com

Permission has been granted for the following:

By Eliot Weinberger, from *Nineteen Ways of Looking at Wang Wei*, copyright © 1987, 2016 by Eliot Weinberger. Reprinted by permission of New Directions Publishing Corp.

From "The Dry Salvages" from *Four Quartets* by T. S. Eliot. Copyright © 1936 by Houghton Mifflin Harcourt Publishing Company, renewed 1964 by T. S. Eliot. Copyright © 1940, 1941, 1942 by T. S. Eliot, renewed 1968, 1969, 1970 by Esme Valerie Eliot. Used by permission of HarperCollins Publishers.

The Library of Congress has cataloged the hardcover edition as follows:

Names: House, Patrick, author.
Title: Nineteen ways of looking at consciousness / Patrick House.
Description: First edition. | New York : St. Martin's Press, [2022]
Identifiers: LCCN 2022019286 | ISBN 9781250151179 (hardcover) |
    ISBN 9781250151186 (ebook)
Subjects: LCSH: Consciousness.
Classification: LCC BF311 .H6526 2022 | DDC 153—dc23/eng/
    20220614
LC record available at https://lccn.loc.gov/2022019286

ISBN 978-1-250-15119-3 (trade paperback)

Our books may be purchased in bulk for promotional, educational, or business use. Please contact your local bookseller or the Macmillan Corporate and Premium Sales Department at 1-800-221-7945, extension 5442, or by email at MacmillanSpecialMarkets@macmillan.com.

First St. Martin's Griffin Edition: 2024

10 9 8 7 6 5 4 3 2 1

To Claire Roscow (1988–2010),
Dr. Jonathan Leong (1983–2019),
and Cam Christie (1982–2020);
all, whose neurons turned in early

Contents

The movements of animals may be compared with those of automatic puppets, which are set going on the occasion of a tiny movement; the levers are released and strike the twisted strings against one another.

—Aristotle, "On the Motion of Animals"

Nineteen Ways

of

Looking at
Consciousness

Introduction

If I were asked to create, from scratch and under duress, a universal mechanism for passing consciousness from parent to child, I would probably come up with something a bit like grafting a plant. Each parent would donate a small piece of their brain and place it on some sort of growth medium, maybe some agar, or some flour with sugar and yeast, and the child would sort of just expand, like those water-absorbent foam dinosaur toys, into its final shape around the pieces of parental brain until it, too, was conscious. How else could it possibly work?

Instead, something much more remarkable happens in nature. An entirely new creature can grow into a fully conscious version of itself, and the entire process occurs, as if by fiat, anytime a certain kind of single cell with the right mix of nucleotide sugars is kept alive for long enough. Which means that consciousness is not something passed on or recycled—like single molecules of water, which are retained as they move about the earth as ice, water, or dew—from one living creature to the next.

Instead, consciousness can be grown from "scratch" with only a few well-timed molecular parts and plans laid out. It is not drawn from a recycled tap of special kinds of cells or dredged from a vein of free will. No, the darn thing just grows. From its own rules. All by itself. And we have no idea how or why.

Despite this, most of the time, except to the philosophers and neuroscientists, there is no "problem" of consciousness. It seems to work just fine, almost effortlessly, which is a large part of its rarely questioned charm. One can have a full and vivid life without giving a second thought to the makings of the first. And yet, even though everybody knows what consciousness feels like, it remains central to the greatest unsolved scientific and existential mysteries we will probably ever face: What is it? And how did it get here?

This book is a collection of possible mechanisms, histories, observations, data, and theories of consciousness told nineteen different ways, as translations of a few moments described in a one-page scientific paper in *Nature,* published in 1998, titled "Electric Current Stimulates Laughter." The idea is an homage to a short book of poetry and criticism, *Nineteen Ways of Looking at Wang Wei,* which takes the poem "Deer Park" by Wang Wei and analyzes nineteen different translations of it in the centuries since it was originally written.

Generally, though not in every case, each of the nineteen "ways" in this book promotes the idea that the brain evolved for, is itself, and will always be dedicated to movement. The best reason to start there is the simple, physical, and verifiable fact that a human brain's only outputs are the muscles it connects to, whether the small ones that dart the eyes back and forth or the lumbering thigh muscles that kick-start a walk. No matter what goes into a brain, only movement ever comes out.

And though I believe that this is the correct starting point for understanding consciousness—that it can be fully understood from the brain's parts and their goals—this starting point is by no means universally agreed upon. There are as many possible lenses to translate the details through as there are ways to hold a lens. Here, every chapter points toward an explanation for a few moments during the neurosurgery described in the *Nature* article, when a still-awake patient had her brain carefully prodded by an electrode capable of producing blasts of electricity and altering both her behavior and, arguably, her consciousness. Throughout this book, I will call the anonymous patient Anna. (Only her initials, A.K., are given in the original article.)

As a whole, the arguments and ideas in this book represent some of the most popular or plausible theories of how consciousness arises, works, feels, or degrades with use or error. Each chapter does not align neatly with any one theory or thinker, and some theories find their way into multiple chapters. In part, this is necessary because the many theories of consciousness today are as varied as the scientific fields they draw from, a funny quirk of human reason. For those studying the origins of life on the seafloor, for example, consciousness may be the natural interactions of pH, proton pumps, and evolved metabolic efficiency over time. But another theory, from someone who happens to know psychology and decision-making well, may be explained by lots of org-chart-style boxes and arrows. To those scientists who study attention, perhaps consciousness is the mathematical equivalent of a spotlight; to those who study eye movements, it might be the grand effort to cancel out self-generated movement; to those who study flies, it could be as probabilistic as a population of midges' decision to go left or right or loop back on itself; to those who study songbirds, it might be the cognitive consequence of the rewiring of language-learning loops; and so on. There are, in other words, almost as many theories as there are thinkers out there.

I can be clear about my biases, too. I spent the better part of a decade doing laboratory research on a tiny parasite that infects mouse brains and maybe makes the infected host mouse prefer the smell of cat urine just a little bit more than they had before. Thus, I am a mind-control parasite guy. I think in host-parasite terms and relations. I think of free will as something like a river's current with a natural direction and kind of physics to it, which also can, under the right conditions, be redirected. However, as with Wang Wei's original poem, which is lost to history, I believe there is still a way to inch closer to an unknowable truth, guided by the sense that one is on the right track.

For this reason, some chapters are explanations of

consciousness as a single allegory, comparing the evolution of consciousness with the evolution of the game of pinball (chapter 2) or even to a small town (chapter 10). Other chapters base the explanation on the latest scientific research into how a brain controls a body's movement (chapters 3 and 17) or on how consciousness arises from waves of electrical activity (chapter 4), language learning (chapter 12), causal power (chapter 14), or efficient compression of information (chapter 11).

Some chapters seek explanations in the physics of temperature and prediction (chapter 8) or in the quantum realm (chapter 15). Others propose that the best clues are found in the minds of learning robots (chapter 18) or in the mind of a cat with thousands of toes, should she lose some of them (chapter 16).

Others consider why removing parts of the brain changes some aspects of self but not all of them (chapter 13) or how and why a conscious brain is often telling itself fictions (chapter 1). Some consider whether the brain is more like a simulation (chapter 6) or a radio broadcast (chapter 9) with networks in the brain the receiving antennae. Others take issue with the very idea of theorizing about consciousness and defining its terms (chapters 7 and 19) or wonder what answer humanity would give if forced (chapter 5).

Each of the nineteen ways described in this book is a patchwork or tapestry of scientific findings and arguments and a collection of compelling or interesting conclusions organized around a few themes or arguments.

For the sake of clarity, and readability, each translation is self-sufficient, and requires no background knowledge of the philosophical or scientific work on which the chapter is based. Names, credit, and caveats have been excluded from the main text and included in footnotes, which give context, references, and sources as appropriate.

The twentieth and final chapter ties all of them together.

The hardest part about studying consciousness is that it hides by making the outside world into a version of its wishes, trapping its host in a kind of virtual, fragile snow globe we call subjective experience. We are stuck studying it from the inside, and any attempt at explanation is always going to be an attempt at translating kinds of experience.

In *Nineteen Ways of Looking at Wang Wei*, the book's editor directly confronts the problems inherent to translation. I have a similar hope for this book, where contradictions, ambiguities, and debates around relative priority of theories or thinkers are encouraged, rather than dismissed, as they often are in academic discussion around consciousness. Who, for example, has authority to speak about consciousness? Do scientists poking around mouse brains really know more about human brains than therapists or manicurists, who have spent tens of thousands of hours probing why someone had a bad day? Does an electrode really tell you more than a question? (If so, by what measure?) Is there one answer for how all consciousness works across all of life, or one answer per consciousness, per moment? When different explanations are possible, which aspects of which theories get collapsed or lost and why?

In answering these and related questions, this book offers complex arguments and thought experiments that highlight the differences and similarities among many of the major modern theories of consciousness. The hope is that, through repetition and variation from different angles and points of view, a clearer sense of each and, ultimately, of the original—that is, the subjective feeling of what it is like to be conscious—can emerge.

Why another book on consciousness? Why now? When I was training for my Ph.D., at Stanford, I was once asked by a local vacuum repairman if I could help explain to his brother, a drug addict, what addiction truly was down there in the weeds of his brain. When I responded that I could not, that I am just a laboratory researcher and that, besides, we do not know what addiction is because we do not know what brains are and that addiction is social and complicated and context-dependent and its mechanisms mostly unknown, he responded ruefully, but with a kind of haughty resignation, that he works with vacuums because he can fix them.

That same year, a brilliant friend of mine, whom I had met at a neuroscience retreat, died by suicide, at age twenty-two. I was told by another of her friends, before I had even asked, that she had always been "very sad." Neuroscience is a frustrating field to be in. We have mostly failed, as there is little point to celebrate an understanding of small pieces of something if you cannot prevent the failures of its whole. We know very little about the things we wish to know much more about, and the strain on mental, subjective well-being today is, from the global numbers, only getting worse.

Sometimes, I imagine that modern neuroscientists share both the wonders and frustrations of an astronomer in, say, the Babylonian times; the simultaneous wonders and frustrations of looking up at the sky every night in awe at what was and was not known about the motions of the points of light up there so impossibly high in the sky. At the time, Babylonian astronomers knew where stars would be next but not why; they knew that some stars, which were actually planets, would appear to move backward in the sky but, again, not why. Their guesses, and they had a lot of guesses, were all wrong.

Similarly, today, across the many scientific disciplines that study the brain and wish to know its secrets, we know where activity occurs in the brain and can maybe even predict where it might occur if you show someone, say, a face—in the "face area" of the inferior temporal cortex—but we do not know why. I have heard it said that the difference between physics and biology is the difference between Galileo dropping both a bowling ball and a pigeon off the top of the Tower of Pisa. But really the difference is that physics can land an autonomous robot on Mars with pinpoint accuracy, study the origins of the Big Bang from under a mountain in Italy, and split a literal atom in half to unleash the cosmic hell of a thousand suns, while we over in neuroscience cannot even tell you what being "very sad" is. That gap in knowledge frustrates the search for consciousness. And the longer we go without understanding it, the more people we will lose for inexplicable reasons.

My preferred translation of Wang Wei's poem "Deer Park" is the following literal, character-by-character attempt at translating the twenty ideograms directly from classical Chinese.

Empty	mountain(s)	(negative)	to see	person
	hill(s)			people
But	to hear	person	words	sound
		people	conversation	to echo
To return	bright(ness)	to enter	deep	forest
	shadow(s)			
To return	to shine	green	moss	above
Again	to reflect	blue	lichen	on (top of)
		black		

What I appreciate most about this translation is that many of the individual ideograms have multiple or ambiguous meanings that the translation leaves unresolved. The middle character of the last line can be translated as "green," "blue," or "black"; the fourth character of the last line as either "moss" or "lichen"; the second character on the first line as "mountain" or "mountains" or "hills"; the second character on the third line as "bright" or "brightness" or their opposite, "shadow."

The character-by-character translation does not settle on which version of a concept or word is "correct." In my years of studying the brain, I have found no more compelling analogy to the difficulty of understanding the human mind than this single page of poetry. It has been said that the uniquenesses of humans' consciousness is our ability to hold more than one contradictory idea in our head at the same time, a feat that this translation embraces.

What does it mean to translate consciousness when some people see images in their mind's eye but others, called aphantasics, "see" nothing? What does it mean that there are twenty amino acids that make up all of life on Earth and that, depending on the context (like each ideogram), can take on separate roles? What does it mean that some people have more kinds of light-responsive cells in their retina, which may allow them to discriminate more colors than others, or that some of those same people see dresses as white-and-gold striped and others see the same dress as blue-and-black striped? Or that some people remember things and can replay them just as they lived them while others see memories replay as if they were broadcast on a television one hundred feet away?

What does it mean for the ability to talk about the human mind, as if it is only one thing, that one person (James Joyce) described the inside of his mind as a "grocer's assistant" while another (Albert Einstein) claimed to be able to visually imagine the speed of light as a teenager? Or that when the first description of the "stream of consciousness" was used, to describe a work by the novelist Dorothy Richardson, she strongly objected to the phrase, saying that consciousness to her, instead, "sits stiller than a tree"? Which of all these versions of consciousness is "correct"?

Nobody is wrong, but in a way everyone is, because their version is theirs alone. So, too, must we start at the simplest, most literal translation of consciousness to have any hope of grasping it. The character-by-character translation. The one and only thing we know for certain for every one of us. That there is something that it is like to be us. That's it. Everything else is unknowable excess. One of the beauties, and frustrations, of talking about consciousness is that everyone can know only their slice of the world. We have many tools, like language, gesture, and theory of mind, to try to jump into others' heads, but at the end of the day we can only ever scratch the surface of what really goes on inside.

Some practical notes on reading this book. The original text of "Electrical Current Stimulates Laughter" can be read in the appendix (page 199). It is mostly readable to nonscientists. However, this book is meant to be read with no knowledge of the study in question, and it is explained in pieces, sometimes just as a hint, in every chapter. The basic story is that a neurosurgeon, using small, carefully placed blasts of electricity to the brain, was able to cause the patient, Anna, to laugh. Alone, this is not surprising. We have long known that electricity powers our muscles to act, and laughter is just a series of rapid, coordinated muscle movements. What was so surprising was that Anna said afterward that she also felt the subjective sensations of joy and mirth alongside the laughter and that she, when asked why she laughed, gave different and implausible answers each time.

This book is based on the idea that, like an unsplit atom, Anna's story contains within it hidden multitudes and that any attempt to explain consciousness should be able to fully explain the events of that day, both inside and outside Anna's head. Thus, each chapter title can be preceded by the words "Consciousness is . . ." These titles do not always represent my opinion, nor is each statement inarguable fact, and all are best read with the presumption that the point of view is being argued by someone who believes strongly in that chapter's claims or conclusions.

A clear example of this approach is chapter 7, "The Median Price of a Thrift-Store Bin of Evolutionary Hacks Russian-Dolled into a Watery, Salty Piñata We Call a Head." The chapter starts with a quote from Sydney Brenner, Nobel laureate in genetics and close friend to Francis Crick, that "consciousness" is not the singular problem we think it is and will be remembered in time as a confusing misdirection, like the search for "luminiferous aether" in the late nineteenth century. This was something he told me, in personal communication. I spent days interviewing Brenner, who has since passed away, about his ideas on the brain, in his hotel suite at the Shangri-La, in Singapore. I listened; we argued; I came back the next day.

Brenner, hooked up to oxygen and bound to a wheelchair, had spent each morning combing through genomic databases for the web of life, trawling it for similarity to human genes in hopes of finding one last secret; I, alas, had mostly spent my mornings wondering if it was ironic or unironic that the cuttlefish we ate for dinner the prior night had tasted like nothing. And so, that chapter, its title, and conclusions are my best efforts to summarize his idea that brains are an evolutionary happenstance full of errors and that the lessons of history apply to genomics and neuroscience all the same. It is his translation, in a sense, but in my words. I'm just the ghostwriter, as everyone translating "Deer Park" was for Wang Wei, for millennia hence.

Similarly, other chapters do the same for a variety

of thinkers and theories, some of whom I have met or interviewed for this book and some of whom I have followed and read from afar. (There are many theories and thinkers that this book owes a great debt to—sources, notes, references, and further reading can be found in the Notes section at the end of the book.)

Any repetition or small variation in repetition is on purpose. Of the fifteen translations of "Deer Park" that are in prose, the description of "moss" or "lichen" is translated thirteen times as "green," once as "blue," and twice as no color at all. What a reader is left with if they developed all the translations, as a kind of photograph, would be a single image, multiply exposed, with a combination of mostly green, a little bit of blue, and a little bit of nothing. I believe that every book on consciousness is confronting a similar translation effort, even if naively.

Ideally, any explanatory gaps along the way ("Wait, how did we get from microtubules to the subjective feeling of joy, exactly?") also exist in nature. We are pretelescope. Pre–Isaac Newton. Some of the biggest and best brain-reading devices in existence have not been the clear keyhole into the hallowed rooms of the mind that we had hoped for decades ago. Instead, today's tools are more like the lens grinders that laid the foundations for Galileo to, one day, make glass smooth enough to curve and capture just enough light for him to turn a telescope upward and outward.

One last note. In any act of selection, people can get sensitive about rankings or inclusion. If you know of or adhere to a particular theory of consciousness that does not get mention in this book, and take umbrage, I defer to a line from the editor of *Nineteen Ways of Looking at Wang Wei,* which both addresses and immediately closes the door on a similar problem, but for poems: "I have presented only those definitions that are possible for this text. There are others."

Whatever Wang Wei or evolution intended for their respective creations no longer matters. All we have left are versions. I have presented only those theories that I believe give the best clues to explaining the mysteries, horrors, and awe-inspiring details of consciousness and its multiple explanations. (There are others.) At the very least, the various and sometimes contradictory theories are essential background to the gathering of stable and accurate observations about brains and minds, which will persist as data for future scientists or philosophers to figure out what it all means. There is a precedent. For many centuries, observations about magnetism and electricity were tallied and noted, but it was not until a theory came along relating the two that the cataloged unexplainable phenomena were suddenly understood as the symptoms of fully explainable interactions.

Perhaps the same will happen, one day, with consciousness.

Relative to the Observer
Who Is Also a Liar

In the mid-1990s, Anna, age sixteen, had brain surgery to alleviate suffering from epilepsy. Because the brain does not have pain receptors,[1] she was allowed to remain awake the whole time, with only a tiny amount of localized anesthetic to numb her scalp. During the procedure, the doctors and attendants asked Anna a series of questions meant to keep her talking and, as they probed her brain with electricity and micron-thin blades, hoped that she would not stop.

Though the language parts of her brain were roughly in the same places as they are in all other human brains, the brain moves with each pulse of blood, and every brain, like every coastline, has its own slightly different contours. If Anna had stopped speaking at a certain spot of stimulation—because the electrical current could activate cells relevant to thought and speech—the surgery team knew they were in an area important to language and thus one to be avoided by the surgeon's scalpel.

Oddly, they did not ask Anna while her brain was being probed with electricity to write poetry and stop when the poetry became bad. They did not ask her to intuit a response to a fictional domestic dispute and stop when her response became improbable or immoral. They did not ask her how far in front of her eyes her visual imagination extended and stop when the distance became too uncomfortably far or the description too Daliesque or she suddenly lost perspective. Instead, they asked her to do a variety of seemingly mundane tasks, including name objects, read, count, and flex her hands and toes.

Many imagine the human brain as a series of lit-up wires connected together like telephone poles, strung inside a snow globe made of bone, each neuron brightening like a candle when it has something big to say or wants its owner to notice something. It is not. There is no light, for one. The brain is messy and venous and dense and soaking wet, all the time, and is about as heavy as a hardback copy of *Infinite Jest*.[2] It is not designed, perfected, or neat. It is a thrift-store bin of evolutionary hacks Russian-dolled into a watery, salty piñata we call a head.

If the surgeons had poked Anna's brain with the tips of their fingers, which they would never do, but surely has been done, her brain would have had the give of a very soft Brie cheese. The surgeons could thread a small, loose wire straight through it, but never would because along those tracks could be memory, identity, and bits and pieces of the girl's sense of her teenage self, which is to say the accumulation of her preferences. Hanging like a furled sail off Theseus's ship,[3] a surprisingly tensile outer shell called the dura mater (Latin, and Freudian, for "tough mother") would be visible near the girl's head, as seen by those in the room, though not by Anna, who just wants her seizures to stop.

The surgeons eventually found a spot on her brain while operating that, when stimulated, caused her to laugh, a discomfiting sound in any operating theater. More technically, the surgeons used their fancy electric wand to produce an electrical current that, because the brain also uses electricity to communicate many of its own messages, caused certain neurons in her cortex to send a signal mostly indistinguishable from the natural one to parts of her muscles to coordinate action of these muscles, and it was these contractions that bounced air between them to produce a sound perceived by those others in the room as "laughter."

Strangely, when asked the source of her laughter, Anna gave a different answer each time. The answer was dependent on her immediate surroundings, and often involved an aspect of a picture she was viewing or a person near her ("the horse is funny," "you guys are just so funny…standing around"), even though the correct answer, involving the surgeon's electrode, eluded her. Instead, she confabulated the reasons behind the laughter and mirth because the brain abhors a story vacuum and because the mammalian brain is a pattern-recognizing monster, a briny sac full of trillions of coincidence detectors that are only useful if there are connections between things. Even a wrong pattern, a guess, is at least a pattern to learn against.

Though she did not receive general anesthesia, those before and after her in the same operating room did, and a full description of her awake brain must also explain one of the most remarkable things about consciousness. It can be silenced by anesthesia, in part or in whole, only to recover fully again in a few seconds, minutes, hours, or days. That there is no one kind of anesthesia to turn the consciousness dial to zero means that, although all conscious brains may be alike, the conscious brain has many, and sensitive, failure modes. Consciousness, like Tolstoy's unhappy family, has only one way of adding up to a whole, but many ways of falling apart: xenon gas, propofol, isoflurane, cocaine, nitrous oxide, barbiturates, benzodiazepines, and ketamine, each with different chemical profiles and causes, can all silence consciousness.

Some anesthetics, like propofol, which is sometimes called the "milk of anesthesia" because it is white and oily and repels water, can cause bizarre effects. People who were crying before anesthesia came out of it, hours later, again (still?) crying. These effects are less like a pause button for consciousness and more like a needle lifted off a spinning record.[4] When waking up from anesthesia, or coma, a state sometimes called post-trauma amnesia, people will often have strange, lewd, or primal behaviors, speech, or urges. Legally and socially, people are not often held responsible for what they do or say during this time, which makes one wonder why we ever are. Consciousness is all and every one of these states equally. Any good theory must fully account for each of them.

That consciousness disappears nightly is another of its quirks.[5] Though nobody quite understands what sleep is, we know what it looks like and that anesthesia does not induce it. Some animals can sleep with only half their brains at a time, allowing basic functions of consciousness to persist so that they don't fall out of the sky or get eaten. Those in the water who don't fear being eaten, like humpback whales, often sleep vertically, often in groups, like the large towers of an aquatic city, for less than ten percent of their day. Sleep concerns are highly specific: birds dream of bird problems, whales of whale problems, dogs of dog problems.

Those that can lucid dream, which is a kind of awakeness within dreaming, or an awareness that one is dreaming, can be trained to move their eyes while in the lucid state, under rapid eye movement, and these movements under the eyelids can be detected by an infrared camera.[6] Interestingly, this means one can make a code, like those in a video game or medieval monastery, that lets one break the subjective fourth wall and communicate with the great sleep researchers in the sky. For example, a person can learn to, if lucid, move their eyes in a certain pattern and then count to ten, after which they move their eyes in that same pattern again, to mark the end of their test. Remarkably, some people take around ten "objective" seconds to do so, which implies that their subjective, incepted time—the waking dream within the dream—not only has a time keeping device but that it may be the same one we always use.

Where did this conscious, clock-making observer, lucid or otherwise, who can wake up and take in their immediate surroundings—who, in physical terms, the speed of light sticks to, who only in observation can collapse a wave of light into a position and determine whether Schrödinger's cat is dead or alive—come from?

From the oceans, of course. The most useful thing that land offers that the oceans do not is the large distance at which things can be detected. Visually, swimming in water is like driving in a milky fog, which reduces the range of even the best mammalian eyes. A small bacteria can sense, crudely, in a small shape around itself—we can call this "sight" if it can respond to a light source, or "smell" if it can detect an unwanted chemical toxin nearby—and the total of its sensory range, the full addition of all its input, stretching through and combined across all its sense, is called its sensorium. The experience of any underwater creature paying attention is an experience of underwater objects or other creatures popping into frame at such high speed that an underwater sensorium needs timing closer to reflexes than contemplation. Even in the clearest water, light scatters and degrades over only a few meters, which means there isn't much need for a brain to come up with long-term planning, because what would be the use?

Thus there isn't much need for clock making beyond intervals of a few seconds, which means that there is no need for the brain to whir up an emergency motor-response plan for the shark cresting over the horizon of the Adriatic shelf, because there is no way to sense the shark cresting over the horizon of the Adriatic shelf. This usefully constrains the metabolism needed to keep track of the far outer radii of the outside environment and means any need to plan movements is limited to the timing of events with a small, near-reflexive range.

Hiding in the center of these plans is the observer, the conscious creature, who is just an accumulation of movement preferences and plans trapped inside a sensorium, keeping track of what it thinks the objects around it are and what it might otherwise do with itself.

The great move of life onto land from the milky oceans changed the range of timing that the newly landed needed to care about. The expanded range of being able to see farther through the crisp air, and with new eyes to boot, meant that prey vis-à-vis their predators had to plan to move themselves across alternately sparse and cluttered landscapes in order to find and not be food. To plan, one needed a sense of time, in order for there to be something unto which the plans unfurled; for there to be a sense of time, there had to be a timekeeper. Thus the timing that mattered most for sensing, predicting, and planning depended on a creature's sensorium but the exigencies of land, like gravity and tripping, made it suddenly necessary to plan seconds and minutes ahead.

In a kind of spherical symmetry, minutes of forward planning (imagination) required minutes of backward recall (memory) and, like an inflating balloon expanding evenly and temporally on all sides, landed creatures needed to be able to pay attention to the future and in so doing pay attention to the past by the same amount. To know what a lion cresting the African horizon will do, one must be able to keep track of what similar-looking creatures once did after cresting similar-looking hills.

Mammals like dolphins or whales, which crawled back into water after a brief stint as hippos, saddened by how murky it was, used all the vocal tricks learned on land to re-create through echolocation and sonar, as much as possible, the range of visual distance the eyes granted.[7] Because some sound waves travel through parts of the ocean almost as far as light waves travel through air, and because a brain can be thought of as a tool, like any other, to make sense of and expand an animal's sensorium and to efficiently use the information, it seems clear that aquatic mammals have successfully re-created the benefits of moving onto land.[8]

On land, mammals see to the horizon three miles away, but in the sea, they hear it. On land, cave-dwelling bats and two also-cave-dwelling bird groups evolved echolocation, which is a kind of sonar ability to create sound and understand, from the way it bounces back, what lies ahead. These echolocating land species, most or all of whom dwell in lightless caves, faced a similar visual difficulty in the cave as does life underwater, which proves that the brain, as always, does the best it can with whatever information it is given—sound, light, or touch are just fine, if it is all a brain can get.

At birth, the empty brain knows no stories. Seeing is an experience-based inference performed effortlessly and expertly by the adult human brain. The first time your brain lied to you was the second time you opened your eyes. Teenagers with cataracts, blind since birth, upon opening their eyes after cataract surgery and seeing for the first time experience featureless, depthless, shadowless blobs.[9] They could "see" the same number of photons as an expertly seeing adult but their brain sees nothing in the raw stream of light. Their brains had never seen a coincidence before. They had never walked past a brick wall's corner and noticed the lines of light bend around its edge or watched shadows elongate at dusk or compared, from all angles, sunlight shining through a tree versus the light given off by a tree full of tiny white candles.

All these kinds of stories—the girl's of her laughter, the blind of their first visions, and the newly awakened of their sleep—are how the brain hides its strange workings from its owner. Its apparent effortlessness comes at some cost. After the Italian explorer Marco

Like the Rise and Fall of Pinball

There are somewhere between one and 8.5 billion ways a brain can work, which means there are between one and 8.5 billion factorial ways of looking at consciousness.

One way to think about the evolution of a simple brain into a primate brain is like a power station that had to transition, all without ever once shutting down, from coal stoves to steam turbines to electric wires to nuclear to solar to an AI-powered, fuel-agnostic grid. One could easily imagine how dreadful this would be to maintain, with pneumatic tubes sticking out in the wrong places, control panels leading to nowhere, extant-parts lists, corrosive materials, and software incompatibilities.

A colleague once told me he preferred the analogy of a car, unable to turn off its engine, all while upgrading from a Roman chariot to a Tesla.[1] I prefer a third version of the story: pinball, because pinball machines were forced to evolve into both story and storyteller as, once, the brain did too, en route to consciousness.[2] Both are the result of a series of add-ons and user constraints impossible to plan for at the beginning. As such, modern versions of both have legacy strengths and legacy faults.

Like life, the game of pinball is never won but, instead, can be lost less badly at some times than at others. The threshold for what counts for any person as a sufficiently good score—or, as it was once described, the moment during play "when the sun goes down and the stars come out"—is as subjective as the threshold for any one person to live a good life.[3] Metaphors tend to stick to pinball machines like the gum on their undersides because every game, also like life, has so much that feels like chance but isn't and so much that feels like the opposite of chance but also isn't.

The last universal common ancestor, or LUCA, of all modern pinball machines can be traced to 1871, to a British inventor, Montague Redgrave, who was granted U.S. patent 115,357 for "Improvements in Bagatelle."[4] Bagatelle originated in France, in 1777, at a party thrown for Louis XVI at Château de Bagatelle; it can mean "a thing of little importance," "a very easy task," or "a game in which small balls are hit and then allowed to roll down a sloping board on which there are holes, each numbered with the score achieved if a ball goes into it, with pins acting as obstructions." Redgrave, in his improvements to bagatelle, evolved the game by adding a spring plunger, reducing the size of the ball to a marble, and inclining the field into which the ball is thrust.

Early-twentieth-century pinball games that evolved

from bagatelle had none of the modern trappings of to-day's pinball machines, such as flippers, coins, or legs. They sat atop a desk, like large liquor barrels, and one changed the course of the ball as it sloped downward in three distinct ways, as one also does the course of history—slightly, by nudging it; heavily, with all one's weight; or accidentally, while attempting to do other things. To hinder people from simply picking up the machine and modifying the ball's trajectory, designers added ungainly legs in the early 1930s and made the machines heavier, which only disadvantaged the weaker or less leveraged players.

And so, in 1934, a tilt mechanism was introduced to these still proto-pinball machines that prevented the player from moving the machine more than a set amount. Machines could also, for the first time, plug into electrical outlets, allowing them to produce lights and sounds and compete with the sensorium dazzles of motion pictures and World's Fairs, which were still a thing.

The flipper as we know it today was introduced, in 1947, with the game *Humpty Dumpty*, which had three flippers to a side and which, like the development of multicellularity, the human hand, and the atomic bomb, changed the world overnight. Everything preflipper looked instantly vintage. Flippers turned what had been a game mostly of chance into something that one could be good or bad at, compete at, wager on, fight about, cry over. *Humpty Dumpty* looked, to the American consumer, how the bow and arrow must have looked to an Ice Age man after seven hundred thousand years of variations on the hand ax. The "flipper bumper" introduced a way to control Brownian chaos; it meant, contra Newton, that entropy could be slowed and maybe even reversed. The ball had been exclusively a downward-trending thing, but now it could rise, like America, from the ashes of the worst world war man had seen. Unlike death and taxes, the steel ball's demise was no longer inevitable. Protestant America was introduced, again, to reincarnation.

By the 1950s, there were two major aesthetic and functional styles for the machines, which divided players into those who preferred the symmetric machines and those who preferred asymmetric ones. These early machines were much slower than modern pinball, less concerned with points, and were mostly concerned with the completion of small, tactical errands that required step-by-step precision and nonrandom sequences. When solid-state, muted digital computer boards made their way into pinball cabinets, the designers added back the sounds of the clinks, ratcheting gears, bells, and whistles to satisfy the nostalgia gene for the sounds of a writing-on-the-wall, bygone analog era. Slowly, the game became less about accomplishing a goal and more about the accumulation of arbitrary, outsize points. (Even today, some pinball machines reward points simply for playing, as if rewarding the act of inserting a quarter or pressing START alone, even if a single flipper is never hit.)

As the games became more and more about points and stringing those points into sequences of more points, there was a creative lull in the industry. Everything had been tried at least once and, even more damning, the world outside pinball was becoming a lot more interesting and interactive. Hollywood had just had its golden decade, with *Star Wars, 2001, Jaws, Apocalypse Now,* and *The Godfather.* Video games like *Pac-Man, Missile Command,* and *Frogger* were mainstream, cheap, social, and had mapped the goals of the species—avoiding predators, survival, crossing the road—onto buttons and joysticks.

And so, in 1986, pinball made one of its final, and riskiest, gambits: it became fiction. The game *High Speed* introduced a narrative arc and was an instant hit. As easily as most infants know how to grab and suckle, players immediately understood on an instinctive, motor level that the goals of *High Speed* were to change a stoplight from green to red, run the light, and flee from the police who gave immediate chase. Suddenly, the ball was not just a ball but a sports car. The player no longer saw, in the oily reflection of the glass top, their own face but rather the faces of Bonnie or Clyde. They were, for the first time, not playing *with* the small, metal ball but *as* it.

The entire narrative capacity of the human brain to find story in inanimate objects was suddenly brought to bear with every quarter. In the 1940s, psychologists Fritz Heider and Marianne Simmel made a short animation where simple geometric shapes like triangles, lines, and circles of different sizes would bounce around the screen and occasionally clump, bounce off each other, or follow the other shapes around.[5] When viewers were asked to describe what they saw, they described creatures in conflict and told the tale of the shapes with genders, villains, emotions, and moral feats of high heroism in classic story and character arcs. Of course, these conclusions are illusory, as illusory as the stories we tell ourselves about *why* we laugh after doing so, even if it's because a surgeon has an electrode in our brains making it all happen. *High Speed* is to the Heider-Simmel illusion what fentanyl is to morphine and became, in its potency, more of a threat to the belief industry than the gambling industry, because what was happening in its players' brains resembled animism more than entertainment.

The flipper bumpers that at first only delayed *Humpty Dumpty*'s inevitable fall were now gas pedals; the player became the driver of a car the size of a pinball cabinet; the clean execution of a series of precise shots was the clean execution of thought, plan, and action, all while on the lam. The physiological RPM of the inner circulatory whorls and loops of hormones and multiball reward, tens of thousands of times more complex than the game itself, started whirring up in both the escaping player and those accomplices who stood watching, cheering, and abetting. A drain was no longer simply a lost quarter and a reset of points. It was the player's survival and, for the first time since Louis XVI threatened beheading unless his courtesans told him why the bagatelle ball was like the enemies of the French Republic, the game became a kind of storytelling guide to chapter and life. Thus the 1980s and early 1990s became an era of near-Cambrian explosiveness and introduced, through narrative and the miniaturization of circuit boards, the peak of pinball creativity, interest, revenue, and popularity.

In the mid-1990s, however, when both war and actual car chases could be televised live and other kinds of entertainment had become cheaper, more social, and took up less physical space, pinball was again in trouble. A game of virtual pinball became one of the bestselling games on early personal computers and yet, through emulation and software alone, the game kept all of the physics of weight, tilt, and acoustic echo that defined the earlier, realer machines. A series of speciation events on the corporate side of things caused revenue to decline and manufacturers to shutter until, in 1997, there was an official industry bottleneck. Only two major manufacturers remained.

In the late 1990s, forced to evolve or die, there was a push to incorporate the technology of fully digital arcade games that could project light onto reflective screens to give illusions like depth, texture, and horizon. One manufacturer decided to hedge, stealing both the idea and execution of lateral gene transfer from bacteria in order to cobble the best parts of video games and add them to pinball. Just as pinball had changed multiple times before in its history—legs, tilt, electricity, narrative—it would be forced, in a natural process of change and progress, to do so once again. A reflective mirror, computer monitor, and projector were placed over the standard pinball machine to create a chimeric centaur of old and new, which meant the display was no longer simply a passive addition to the playfield. It was a spectacular technical feat because the projected overlay, through a combination of sensors and illusion, seemed to interact with the ball as if by holographic magic. There was no longer a difference between the screen and the playing field or the analog and the digital. It was all a unified thing and could be glanced at easily without a break in attention as if through a single, cyclopean aperture. Bagatelle had become unrecognizable to all but the ancient geneticists.

This late-stage tack, however, was too expensive and too late. One company, whose slot machine and hospitality revenue had steadily increased along with the popularity and broader legalization of gambling, switched to making slots alone.

One of the most beguiling things about the human brain itself is that it, too, is an overlay of the history of analog and digital.[6] The chemistry of molecules moving rapidly within and between neurons had to become, at some point in the evolution of life, the movement of electrical current. Then, bored, because predatory consumers demanded that change be gradual, that feature sets stay across generations, and that the best ideas be stolen and made into a new whole, life added bilateral symmetry, legs, reward circuits with arbitrarily dopaminergic counters, and a vestibular system that, in coordination with tilt circuits in the brain, kept us at an even keel. The analog, single-celled stuff is all still there—much as you can find, if you look hard enough, traces of bagatelle in modern pinball.

And then, at some point, when the competition was too great, and the odds looked dire, we, too, became fiction.

We used paddles and flippers to grasp and swim and carve and cook, and the metal ball became the storytelling self that moved through the world picking up points and victories. If there were time and resources to spare, the goal was to multiply and thus create, from only the insertion of carbon, water, and oxygen into the coin slot, the cyclopean aperture of story projection over the historical tabletop of parlor fancy. We got a consciousness—the projection atop a cabinet full of mechanical and analog bells and whistles that would otherwise do fine enough but are enhanced by the holography—and one that might, in a last-gasp effort at survival and relevance, one day also be fully virtual.

Why did evolution not keep single cells single? For the same reason pinball does not stay bagatelle. Why did pinball not stay mechanical or convert entirely to digital? For the same reason why the signals sent around the outer lanes of our brain are both part chemical and part electrical and why zapping a brain in a particular place, with electricity, can cause feelings of joy and mirth reliant on chemical concentrations: because there was a history to consider, because nobody thought to do otherwise, and because nobody came along and pulled the plug on the mammalian brain, ever.[7]

There is something strange about placing a quarter into a pinball machine and knowing that you will receive nothing physical in return. Those who recoil at the logic and waft of watching sunrise through cigarette smoke in casinos or those who judge others who place quarters into machines for the chance of more quarters but do not similarly judge pinheads are, evolutionarily speaking, irrational. By Darwinian standards, the gamblers at least have a chance, no matter how low, to gain more resources than they initially gave.

What reward is granted to the pinhead except the proxy hormones of victory, delayed defeat, high score, training data, or the chance at a free game? The surest way to never lose at pinball, after all, is to never play.

The Anxiety Felt While Prevented from Migrating

In accordance with their inherited calendars, birds get an urge to move.

—**William Fiennes,** ***The Snow Geese***

A friend of mine, a bird-watcher, once told me that the best time to search for birds is right after a storm because the grounded ones are very anxious to get going again. He called it *Zugunruhe*, a German term, and translated it roughly, perhaps poetically, as "the anxiety felt by migratory birds prevented from migrating." A body, too, is restless to get moving; in fact, the entire purpose of the brain is to make efficient movement from experience, and everything else, including consciousness, is downstream of these efforts.

In Isaac Asimov's 1955 short story "The Singing Bell," a detective is trying to solve a tricky case, a kind of murder-mystery variant of the two-body problem, from physics. The case: someone had stolen, and possibly hidden, a rare object about the size and shape of a small harp, called a "singing bell," from the moon.

Back on Earth, the main suspect, a master thief, denied involvement. There was little to no physical evidence against him, no digital surveillance or GPS tracking (which didn't exist at the time), no obvious biomarkers of space travel caused by radiation or the shortening of telomeres. Further, there were no witnesses and the known thief had a decent alibi—paper records showed that he vacationed that summer, as he did every summer, at the exact period of time during which the crime was committed. Look at the pattern, cried the suspect. The detective, stumped, asked a scientist who owned one of the rare singing bells to help out. Together, they invited the suspect to the scientist's home, where the scientist casually asked if the suspect would toss him one of the singing bells. All in the room watched, aghast, as the toss fell far short. The singing bell shattered on the ground.

Et voilà, said the scientist.

The only witness to the crime, the suspect's unconscious, had just cosigned their confession. The suspect had tossed the bell as if it were much lighter than it should have been, as if it had the mass determined by the lesser gravity of the moon. In other words, the expected dynamics of the toss were embedded in his brain based on his experience on the moon, and the earth's gravitational pull gave him away. On the moon, a part of his brain kept a record of exactly how heavy the bell was and thus all the possible forces that might be required to, one day, toss it. One cannot hide from the statistics of what these cells know. One cannot hide from the neuronal weights and the need for prediction of the future from past experience.

The story's resolution hinges on a key truth about all brains: they carry with them, in their assumptions and lessons, statistics about the world they act in. That all of our brains have two different visual pathways, like off-ramps, to independently keep track of "where" something is and "what" something is, offers a clue about both brains and the world they do their best to understand. This strange anatomical clue hints that, in the outside world, the same object can exist at different places. This sounds obvious, because to us it is, but just because it is a fact of *our* universe does not mean it had to be. (One could imagine a universe where objects somehow changed every time they moved—there would be no need for a "what" or "where" pathway, in such a world, because knowing both "what" and "where" would be redundant.)

In a sense, all movements give away their owner's secrets. Consider the simple act of laughing without reason. When Anna, a teenage girl suffering from epilepsy, underwent surgery to remove the part of her brain that caused her seizures, she laughed after a surgeon stimulated a certain part of her brain. But why did the surgeon pulling a single puppet string *there* in her brain create a whole cascade of muscle activity and emotions? How did one burst of current create so much activity? In part, this was because some actions are so precise, and so rote—walking, laughing—that when the brain activates the patterns of muscular activity, the whole thing goes off at once, like a sprung mousetrap.

But we can go back further, too.[1] The timing of her laugh all started in her mother's womb, before any of Anna's neurons had matched to their destination, and when all her muscle cells had a natural pulse of about ten times per second.[2] At that time, all muscles would fire together, coupled electrically, like rhythmic fireflies. When these muscles later started connecting to nerve cells, two things happened. First, the nerves brought the timing of the natural muscle rhythm into the brain, encoding the timing of these pulses (ten times per second, or 10 Hz) as a sacrosanct feature of interacting with the world. Second, the muscles, freed now by their attached puppet strings, could move independently and start to learn.

Then the brain kicked in, its only job to reduce the metabolic expense of any and all of Anna's present or future movements. Her brain fed off the statistical regularity of observation, prediction, and action because the only output from all of its work are cells in charge of telling her muscles when to move. As Anna grew, parts of her brain, in an effort to reduce the expense of movement, started learning what features of the world might change given certain motions. As a baby, she dropped things off the table, listened to the codes from the cradle,[3] learned what made her mother sad or smile, and watched as her eye movements changed what she saw. The problem that Anna's primate brain faced was that efficient, graceful control of the human hand, with its remarkable degrees of freedom and the near-infinite ways that it and its joints might reach for something, is far too complicated a problem for even her exquisite primate brain to keep track of every millisecond of every day.[4] (Multiplied, also, by the body's approximately six hundred and fifty muscles.)

Thus, the pulses of those early muscles, once they joined to their growing nerves, created a physical limit to their use (again, 10 Hz) that greatly reduces the complexity of coordinated movements. Instead of having to deal with all the body's muscles, positions, and forces at all times, the brain needs to do so only ten times per second—glacial, from a neuron's perspective, which can spark and then slowly count to ninety-nine with the time it has left over. This approximate rate of firing became, in Anna's nervous system, a natural law.

Because the only goal of consciousness is the simplification of movements, as more coordination is required, more consciousness is required. It is no coincidence that some of the more complicated motor feats on Earth— human reach, orangutan swings, bat echolocations, elephant trunk sniffs, dolphin hunts, octopus ambushes— are controlled by some of the more complicated nervous systems. As an adult, Anna would never be able to move her muscles more than ten times per second. If she held her hand steady, this was not the absence of muscular movement but rather a serious effort of the exact kind of movement to suppress the natural, ten-times-per-second tremor that all adult human muscles have. This is all a legacy of the early, fetal bursts. All of us, like the restless birds, are anxious to get moving again.

Any act of thinking is just pretending to act out. Consciousness requires cells that want to move and that know roughly what will happen when they do, but are prevented from doing so. Just as drinking coffee is just running in place; as a mammalian inhale is just moving some cells forward in order to push air along a membrane while standing still; as modern mammalian reproduction is just the process of laying and fertilizing eggs in a makeshift ocean,[5] also while still—so, too, is thinking just moving without motion. Consciousness is the consequence of the primitive irritability of single cells that all share the ability to be impinged upon, to be excited, or to be provoked. The eyes dart back and forth to see; the ears cock to listen; the nose sniffs to learn what it can. Our senses are not passive. They are actively searching the world.

We are irritable animals that move well and think no faster than ten times per second;[6] we are the product of the reduction of the one-trillion-body problem to the one-body problem; we are the product of a brain that is creating hypotheses the entire time about how best to act, knowing that all the outside world can do is irritate

it and all it can do is learn, while alive, as much as possible from these irritations.

Anna, immobile and in the operating room, prevented
from migrating, nonetheless took flight in laughter.

4

The Music While the Music Lasts

> *. . . music heard so*
> *deeply*
> *That it is not heard at all, but you are the*
> *music*
> *While the music lasts.*[1]
>
> **—T. S. Eliot**

L ike the oceans, like an orchestra, the brain has cadence. Consciousness is a result of the neural beats, called oscillations, that coordinate cellular activity across its surface and deeply across its poles.[2]

We may think that sharks have it rough, having to move forward to breathe, but mammals, too, must inhale. We simply took the whole-body motion required for oceanic species to get oxygen and internalized it, as lungs, evolved from a gas-filled bladder or maybe the organ for buoyancy in fish, and hooked it up to muscles that could inflate and constrict to achieve motion while the rest of the body stayed perfectly still, as if on a treadmill. Similarly, we took the motion required to predict the effects of our motions and internalized them in order to achieve thinking, also possible while perfectly still, as if on one of the Red Queen's treadmills, from Lewis Carroll's *Through the Looking-Glass*: "Now, *here*, you see, it takes all the running you can do, to keep in the same place."

In order to keep track of any such motion, all nervous systems in all species must somehow keep track of timings and timing mechanisms to reflect both the outside world's changes and a brain's own. Some of these timings are learned at a species, or clade, level; some in utero; and some, like jet lag, only days later.[3]

Consciousness is present in all timekeeping brains because it has to be in order to coordinate asynchronous inputs into any kind of meaningful, purposeful motion. If you touch your toe and nose separately but at the same time, the events feel co-occurring to your brain even though they are, from an absolute perspective, asynchronous. It took a few milliseconds longer for the signal indicating touch to arrive from the toe than from the nose. That the mind cancels out these time delays is likely a feature, and consequence, of the cacophony of noise in a brain at all times. An individual neuron is blind to its surroundings and does not know, or care, whether it is connected at its ends to other neurons or to a muscle and so must treat each the same. From the cell's point of view, a burst is a burst, a flex is a flex, a vesicle is a vesicle.

Imagine a fish, swimming from below the water's

surface of a small glass bowl. As it eats a flake of food, or comes up to gasp at the air, it rips the surface of the water and creates a ripple that will move outward and, with enough energy, all the way to the edges of the bowl. These ripples can fold back on themselves at the edges and gain or collide with other ripples as they travel. Also imagine that the fish, for some mechanical reason not understood by modern science, orients itself vertically and comes up to the surface exactly once every second to create a new, small ripple.

Taken as a whole, the fish and bowl are now a kind of crude clock with one-second increments. Together, they tell time.

Now try to imagine, instead, a bowl with eighty-six *billion* fish, each oriented vertically, with their heads facing up toward the surface. Though each also still rises and falls in one-second intervals (half a second to sink and half a second to float back to the surface) things are now very messy at the water's surface because it takes longer than one second for some of the ripples to get very far and because each of the ripples have their own speed, determined by the laws of physics. As the ripples spread out, as concentric circles, they run into other ripples, coming from all the other directions. Eighty-six billion new ripples *per second* start colliding with the previous eighty-six billion ripples from the previous second. Some of the ripples collapse and some gain crest as they run into friendly or opposing waves. The mathematics describing the surface is untenable. The surface is not ordered. It is a maelstrom of turbulent, saturnalian chaos.

Now, uncouple the rise and fall of the eighty-six billion fish. Each can choose how many times per second it rises to the surface, though some, due to their size or weight, can breach hundreds of times per second and others only tens of times a second. Because of turbulence and drag, it becomes more efficient for an individual fish to rise and fall alongside one's immediate neighbors and so, some start breaching in synchrony, in what look like groups, and at the exact same time. But now these groups create so much collective underwater drag that it becomes, for others at a certain distance, perhaps because of their weight, or individual differences in their hydrodynamics, more efficient to breach at exactly the opposite times as the other, nearby, groupings of fish.

And so, underwater, two groups form that alternate almost perfectly as opposites, in a kind of tick-tock oscillation. At the surface of the bowl's water, when these trillions of brand-new ripples spread and collide, something strange happens. The surface is not chaotic anymore, but ordered noise. Now the bowl and its fish, as a system, can keep time at many different timescales.

Because so many versions of these smaller, oscillating fish groups start appearing naturally all around the pond, the ripples on the surface have a very specific kind of pattern to them, called pink noise.[4] One implication of the pink noise, as if by magic, is that the surface of the water now has memory. As each ripple is shaped not only by small nearby waves but also by all the large faraway waves (which are themselves shaped by and shape their nearby smaller waves), so every ripple in almost every corner of the pond is a consequence of everything that has ever happened to it.

It is impossible in the real, physical world for ripples to precede their cause and likewise for consciousness to precede its causes, which are ripples of electrical activity across the surface of the brain's cells. These electrical ripples sweep back and forth two hundred times per second, collecting the computations of local ripples and binding them into a coherent, unified, conscious experience across the many timescales that consciousness requires.

Around twenty-five years ago, when Anna was in surgery, her skull was opened like the top of a fishbowl to the outside world, and because the brain itself does not have pain receptors, she was awake the whole time. The ripples coursing across and through her brain were active, ordered, and noisy. Though there is only one kind of being awake, there are many ways to be not awake. If she were anesthetized, or asleep, the ripples of her brain would be slow and synchronized, as if a group of powerfully coordinated fish had smacked the surface simultaneously and the created ripples had subsumed and slowed all the other smaller ones. To be unconscious, in other words, is just to have one's timekeeping changed.

The surgeons could have, if they so chose, measured

the *kinds* of ripples that emanated from Anna's brain as one might test a body of water to see if it had frozen over. A measure of the complexity of these ripples as they spread could tell us whether Anna was fully, minimally, or not-at-all conscious.[5] If she were unconscious, the pond would be frozen over and everything slowed down. If she were dead, the water would be filled in and all the fish encased, belly-up. The stranger the ripples and the more they changed as they spread, the more conscious, in some sense, Anna would be. A magnetic pulse dropped into a conscious brain will, like a rock dropped into water, make ripples as the activity spreads outward toward the skull's shoreline.

If she were in some way unconscious, dropping the "rock" would cause no new ripples as it bounced off the surface, languid, until it, too, came to rest.[6] A lot is still happening beneath the surface, even if Anna was unconscious or anesthetized. All the fish working their hardest not just to keep time, but to sharpen their ability to keep kinds of time at different speeds, like the signatures in sheet music that guide the speed of the ripples through the air that we interpret ultimately as being a conscious self in control of our own actions but that really are just the music while the music lasts.

A Secondhand Markov Blanket

To her, I'm just the same guy I always was. I mean that he always was. Were. Whatever.

—Driver: San Francisco

Where does a consciousness end and the rest of the world begin? Where is the line between inside and outside? Between life and not life? Between the parts of the universe that are conscious and those that are not? Between you and not you?

To build up a charge, a gradient, or natural selection, there needs to be some kind of a border, but physics and biology draw their borders differently. (Drop both a pigeon and a bowling ball from a rooftop, for proof.)

In the 1974 film *Dark Star*, an artificial intelligence is taught a few basics of René Descartes's cogito, ergo sum ("I think, therefore I am") arguments and, after realizing that its purpose is simply to explode, the AI proceeds to ignore all further human commands and blows up itself, the ship, and the crew. Likewise, as a thought experiment, let us imagine an AI closer to home and given planet-busting nukes that is taught the basics of existentialism and proceeds to become curious about itself. It may start to wonder about the causal chain at the beginning of what feels to it like its thoughts and, realizing that humans only mobilize when catastrophe is imminent, it might give us an ultimatum:

Dear *H. sapiens*,
You have five years to provide a complete description of free will; or, the exact border of Anna K.'s consciousness while she was in surgery, in Los Angeles, in 1996. Or, I blow up Earth.
Warmly,
The AI

The AI then provides the experimental details. In five years, it says, the AI will put Anna through a random set of subjective and objective trials, states, and tasks, and we, humanity, must be able to give a complete and total rolling prediction of every single one of Anna's thoughts. The AI agrees that, if this is impossible, it will settle for a statistical distribution of probable or highly likely Anna thoughts instead of an exacting list of them all. If both of these prove impossible, because free will is truly free, the AI adds an allowable success condition: it will settle for an exact, atomic description of where Anna's consciousness ends during her surgery, in 1996, as long as it accurately defines the line between Anna and not Anna during the experiment trials.

Most of Earth thus mobilized toward figuring out what is widely thought to be the easiest problem of the three: the line between Anna and not Anna. At first, an Earth-wide census was collected where almost everybody, no matter how wild or speculative, had their opinions heard. Some, the linguists, noticed that the problem was very similar to what the psychologist William James once posed for language. How, in a written sentence, asked James, does one know where the words end and the sentence begins?[1] Perhaps we could prove by analogy, they said, that likewise there are similar borders for brains and consciousnesses if only we could define where the neurons end and the person begins?

Others, the entomologists, noted that we should be able to answer smaller, simpler versions about nature and work our way, so to speak, up. They considered a spider hunting on its web. Does the web count as spider or not spider? The vibrations of the web alert the spider to the existence of something; likewise, we "hear" perturbations in the air that compress airwaves from a focal point far away by virtue of detecting vibrations in the hair cells in our ears.[2] Was what the spider does sensing its web's vibrations so very different from what a primate does with the hair cells of the inner ear in order to listen? Is the air not simply a kind of see-through web, a kind of surface on which vibrations travel and information is gleaned? And so, they argued, if we include the ears and the acoustic sensing apparatus as part of Anna's boundaries, should we not also include the web of the spider? Should we thus not *also* count the electrode that prodded her brain during surgery, since it was able to induce laughter, joy, and mirth no differently than if another part of her brain had done so au naturel?

Why stop there, asked these ideas' detractors, partly enraged. Why not include the trees the web hangs from, too? Or the moon that pulls on tides that evaporate air to rain on the trees to grow the branches from which to hang the web? The Big Bang? Where does one stop?

Others, the ornithologists, asked about gizzard stones of birds, eaten early in some birds' lives and necessary for digestion. Surely, they said, we don't count the gizzard stones as part of the individual bird consciousness, do we? Then perhaps we should remove all the rote mechanical stuff from the brain in our descriptions, like the dumb proton pumps or microtubules that, in isolation, are no more interesting than a gizzard stone. If we are looking for the exact line between Anna and not Anna, said some dualists, going even further, we should remove all the unnecessary mechanical elements of her body and brain and leave only the conscious parts—a bit like sifting for gold, they added. But there would be nothing left if we did that, came a response, from the materialists.

Then the microbiologists came along and asked about the possibility of infection or microbiota in Anna's body.

What if Anna had a parasite in her brain at birth, vertically transmitted from her mother, which nestled inside her neurons, as some are prone to do?[3] Is her consciousness the brain minus the parasite, the brain plus the parasite, or are they a kind of amalgamated mind? We should figure this out beforehand, they argued, just in case. Others, the behaviorists, wondered about the tasks Anna would have to perform during the test. What if the AI made her read a book or watch a film? Would the borders of her consciousness change in the act? The world's stories were thus cataloged into those that have attempted to mimic the interiority of thoughts and that thus might, in some sense, insinuate themselves into Anna if she read them. The works of James Joyce, Virginia Woolf, and Julio Cortázar were analyzed in depth for their effects on the reader; the 1947 film noir *Lady in the Lake,* which was shot almost entirely from the point of view of a detective on the case ("YOU and ROBERT MONTGOMERY solve a murder mystery together!"), and Spike Jonze's *Being John Malkovich* saw surprising resurgences, briefly becoming the most popular films in the world.

Some retired neuroscientists dismissed the idea that low-fidelity film and literature should be considered a boundary worry, claiming that only video games offered the correct feedback loop of action to perception required to create an inside and an outside, and thus a border. A study was done on all those who had played the 2011 first-person video game *Dinner Date,* which has one take the role of a man's impuissant subconscious as he drinks wine while being stood up for dinner. Confusingly, some had remembered their experience as an actual memory of having been once stood up, which meant that the fictional story had ingratiated itself into their autobiographical story of self. If a brain is a prediction engine, some memory researchers argued, then would not even false memories, trapped in or as synapses and that shape the brain's predictions, count as internal to their owner? What if the AI makes Anna play *Dinner Date* and she remembers it as really having happened to her? We should prepare, they warned.

Yet still others, the literary critics, noted that unlike novels, video games and film have never had the second-person perspective, the "you." Cameras in video games often take the point of view of a character's eyes and ears (first person) or take a vantage either above or behind (third person) the character, as a drone's-eye view of things. While true, they said, that players often get a vague out-of-body feeling during games and will duck if Mario's head is about to hit the ceiling or will turn their bodies when their go-kart needs to turn a corner in a race, this does not mean that they *are* Mario or *are* the go-kart. It means merely that they can empathize with or ally their consciousness with virtual objects like Marios or go-karts.

We should only spend time on second-person games, they claimed, but no such game exists, so can we just drop it? But can we be 100 percent sure, came the response from the internet, because wasn't there that viral video shared widely once of someone poking under a fridge with a broomstick and when the mouse comes running up the broomstick toward the camera, remember how a vast majority of people watching on their phones drop their phones in fright? Doesn't this mean that just by viewing the video and holding the phone with their hand they in some sense thought they were "holding" the broomstick? If the brain cannot tell the difference between fiction and reality, will we have to include all fictions, as they are experienced, in order to define the boundary between Anna and not Anna?

This took a while to solve. Eventually, a different set of literary theorists pointed out that there is in fact an arguable case for the second person in a video game, worthy of study. In *Driver: San Francisco*, the main character, after a near-death experience in a car accident, can take over the consciousnesses of other characters in the game. At one point, however, while inhabiting the mind of a secondary character, you as the player find yourself in a car chase and are told to chase *your own* car. At which point, part *Inception*, part *Wings of Desire*, you start to control *your* car but from the point of view of the person chasing you. What if the AI makes Anna play that game, some said? Consciousness extends to that which it controls, and the brain's only output interface with the outside world, after all, are the neurons that connect to its muscles. If we count the electrical puppet strings from her brain coursing down into the hand as a part of Anna, why not the simple electrical circuits of the game's controller? (Why not the surgeon's electrode, also, as it stuck out of her brain during surgery?)

Yet still others, the mathematicians and statisticians, argued that the video game and literature stuff was nonsense. Any proof, they claimed, would have to start with a math-based definition between living and nonliving matter. A drop of oil placed in water must diffuse because it, unlike life, cannot maintain its order.[4] Life, they said, is the opposite of the drop of oil because it does not diffuse and can maintain its order against the drives of the universe toward spread, chaos, and heat death. The drop of oil, on the other hand, like the flame of a candle, has no capacity to keep the outside out or maintain an ordered inside because its borders are porous to the diffusing world and it does not resist the universe's temptations. On the contrary, life does resist decay because it has to, and only here, at this border between life and not life, can we say that the continual nesting of this capacity to fight disorder is the difference between Anna and not Anna.

Anna, thus, is a bundle of statistical drives, not biological drives, which create the separations and boundaries. These statistical boundaries are called "Markov blankets" and can nest, like Russian dolls.[5] All we would need to do, they said, is find the level of description for which of Anna's Markov blankets is the most all-containing—which of her Markov blankets, in other words, contains the most complete set of the others. The relations between Anna's parts, they explained, are similar to family relations in that a person's "family blanket" would be their parents, their children, and any other parents of their children. Instead of family relations, the statistical version of a Markov blanket for, say, a single cell would be at the outer edges of its action and perceptions of the world, which aligns nicely with its cell edges, or its membrane.

Anna, though, a complicated primate, with billions of interconnected cells, each with their own Markov blankets, must have some blanket that is the last and greatest. Some blanket in the exact spot where her inside cannot claim the outside as part of its blanket and the outside cannot claim the inside as part of its. Some blanket that is a rolling process sewn and resewn, that stretches and pulls to adapt to the borders of her actions and perceptions with the timing most relevant to her parts. It is, like literal clockwork, remade every second. Thus, it was claimed, no single answer would suffice. Five years of Anna versus not-Anna boundaries would be five years of rolling answers.

One day, a question from a middle schooler, submitted to a contest looking for fresh ideas, confounded all the world's experts: "What about a caterpillar when it turns into a butterfly? When it is liquid goo during its metamorphosis, is it butterfly or caterpillar or what?"[6]

The implication was immediately recognized. The girl is right, said some developmental neuroscientists, that the brain is physically changing its shape every millisecond of every day or every year of its entire life. It is never in exactly the same state. A child's mind metamorphoses into an adult's mind. It does not simply grow. From a purely biochemical point of view, the human brain thus is always comparable to metamorphic goo. Others, the screenwriters, noticed a nonlinear storytelling problem raised by this conundrum. Markov blankets, they noted, are nested spatially but perhaps they could also be nested *temporally*? People were confused until the parasitologists came along and admitted to thinking along the same lines. After all, they said, a single-celled parasite that has multiple life-cycle stages is better thought of as a kind of über-organism with organ separation across time instead of as an organism jammed into a single body at one point of time. And if that is indeed true for single-celled parasites, there is no reason it is not also true for eighty-six billion single cells inside of brains, which also constantly change and tweak their genetic profiles over time.

Yet others, the monists, argued that all of Anna's internal models of the world should be removed from her mind, since above some level of fidelity her internal models become good enough to count as "outside."[7] If the tasks require that she spend a year growing and walking and crawling in a normal home, for example, and then ask her later how many windows the home had, and if Anna is able to trace back her memories well enough to answer the question, then it means she has a little bit of the house in her through her representation of it. Others, the lawyers and race-car drivers, strongly objected to this, saying that a diorama depiction of a car crash was not the same as an actual car crash and that the curved lines of a boat or the aperture of its sails are design products of water and wind. How could these be subtracted from the boundary of the boat itself? Are we seriously playing with dolls while our species is at stake, they challenged?

In response, the librarians pointed out a scene in *The Hitchhiker's Guide to the Galaxy* where a man inverts his walls and places the bookshelves and all the wall hangings of his living room on the outside of the walls. He then claimed, because books always face in, that while under his house's roof he was standing on a small patch of "outside" while the rest of the world, because the walls were facing that way now, was in fact "inside." Maybe, he said, somewhat impishly, the easiest way to define the border between Anna and not Anna would be to define not Anna as a really small and obvious place so that everything else would be, by definition, Anna?

In the end, humanity came together to settle on a fourteen-word answer. Not because it was proved correct but because of the chance, however slim, that the process of verification and fact-checking would take the AI longer than the expected life span of the universe. If so, the threat, on par now with the inevitable heat death of the universe, could effectively be ignored because entropy would win the day, as it was

A Simulation Starring You

*It may be that we are puppets—puppets controlled by
the strings of society. But at least we are puppets
with perception, with awareness. And perhaps our
awareness is the first step to our liberation.*

—Stanley Milgram

The first time we all tried virtual reality was the second time we opened our eyes. We are in a simulation as both puppet and puppeteer and the literary and scientific questions, not coincidentally, are the same: Is it a first- or second-person simulation? How, if ever, will we know the difference?

In Trondheim, Norway, if you stand with your back to the arctic waters and look up you will see, like a needle-point trying to stitch the sky, the tallest building in the country: Tyholttårnet, an old television station with a rotating restaurant on top. It resembles an insectoid Eiffel Tower, with antennae-like hair and a schlocky, Cold War–era UFO on its highest stories, diners where the alien pilots might have sat.

The inner circle of Tyholttårnet's restaurant rotates at a different speed than the outer rim, which completes a single revolution about every hour and contains all the dining tables and the windows. If, while dining, you take a brief trip to the restaurant's inner circle, you will likely become staggeringly disoriented on the way back to your table. All the paintings on the walls and all the doors and holes into the kitchen and all the golden ratios from architecture class will have changed in relation to everything in the outer circle. It will not feel like being in a new room. It will feel like being in a parallel-universe version of a familiar one. This break in continuity, and the brain's uncomfortable surprise, is a peek behind the curtain of the brain and how, every second of every living moment, it is a virtual and fallible mapmaker.

On a basic level, the illusion induced by Tyholttårnet's rotations is evidence that the brain has some sort of stored-up layout of what it thought the room was.[1] Otherwise, how could it compare? A room is familiar when a brain has been there, somehow—when, among the arrangement of things and objects and spatial ratios, it has kept a vague memory of what was next to what. Indeed, there are particular kinds of cells in the brain that keep track of a body in relation to any room or space that it is in. Some of the cells keep track of precise spots in any arbitrary room or space and groups of those cells overlay a regular, repeating, triangular grid that can dynamically map to any space or room.

These cells seem to activate in any room or space one is in, with similar or similar kinds or maybe exact repeats of activity when one remembers having been in that exact room or even if one daydreams about a future plan to be in it.[2] The invisible tiles appear to scale at different sizes for different animals, but always, within a species, in multiples of around 1.42, or the square root of $2^{\wedge 3}$. The grids can also, as far as we know (because how, really, could we test it?), extend infinitely.

In combination with other, undiscovered uses, the activity of these cells likely creates a universal metric for space within all mammalian brains: a hat rack for the infinity of experience. When the 1.42 grid shifts beneath your feet, like in Tyholttårnet, this shift can be very confusing for a primate brain used to a stable ordering of things. Trees in the African forests did not rotate and change their respective places just because our ancestors stopped looking at them, which is, in part, why the illusion is so confusing and frustrating. Each and every mammalian brain in the room at Tyholttårnet bristles as the honed product of millions of years of tinkering so that the brain can make a mental map of each wall, floor, angle, and object in the room suddenly changes every time its host gets up to pee.

In other words, no matter how boring or familiar a room is and no matter whether you are reading this on a plane you have or have not been on before, your brain is spending gobs of stored metabolism keeping track of the walls and ceilings and their spatial arrangement specifically with respect to you, all so that you perceive a virtual reality with you at its center.

Expanding from this, it means that every single object that we encounter in our daily lives, including people, cars, pets, and our own bodies, is also being impartially or virtually perceived. We have only a few ways of interacting with and sensing matter but these are only a few of the hypothetically possible ways, if only we had the right biological tools. What we see is not the object itself, but an evolved decoding of the parts of that object relevant to survival, not to the truth.[4] A red pill is not red. A blue pill is not blue. They are the same color, which is to say that they are no color at all, to all but our eyes.

What you experience are the brain's deductions about the relevance of objects or a room's layout *to you,* to your survival, and to your goals. A cat one hundred feet away is not actually smaller than a cat nearby at your feet. Your brain simply chooses to depict it as smaller because things farther away are, literally and figuratively, less likely to harm you. We cannot take it all in at once and so are blind to the majority of the physical properties of any object, thing, or person around us.

When we see breaks in the brain's storytelling, we learn something mainly about the storyteller. Every door walked through, every wall to your side, and every ceiling above you is being kept track of by a part of the brain that has its own virtual versions. When Tyholttårnet surprises a brain, it is like a storyteller's stutter or a doorknob's jam: suddenly, you notice. None of this, of course, is conscious or apparent to us, and the invisible grid on which all mammalian perception is laid often goes unnoticed, like tiny beetles to conservationists. But it might just be the key to it all, because these cells are doing more than just keeping track of spatial arrangements of things. The same, or similar, cells respond in some mammals—possibly, likely, in us, too—in respective spatial ways to nonspatial concepts, like musical pitch.[5] One implication is that more abstract concepts in humans, like "rising" and "falling" pitch, may have physiological bases beyond their obvious metaphoric ones; maybe "higher" pitches are placed on the same spatial grid and are, therefore, indistinguishable to the brain from the virtual placement of "higher" places; maybe this is also behind the remarkable effectiveness of "memory palace" tricks, where one seeds spatial memory and imagination with information one wishes to recall, like digits of pi or epic poems.

Access to the brain's store of these virtual spatial layouts may allow, more broadly, for all the kinds of thoughts. Brains are so good at structuring categories and concepts and bringing them to the virtual, conscious desktop because, as with language repurposing the hand-gesture parts of the brain, the brain uses the same blueprints when it can. Brain matter is co-opted when fallow. The reason we see when we dream is so that other parts of the brain do not take over the visual real estate during sleep, when the eyes are quiet and their cortical buildings vacant.[6]

Few other observations about brains and consciousness make the simulation argument more compelling than the fact that, just as in a computer, a simple bit of shunted electricity can cause a brain to compile, run, and display to the conscious clipboard any conceivable experience a brain can have: movements, emotions, sensations, imagined movements, imagined sensations, memories, urges, and so on. During surgery, with the skull open and the brain exposed, small bits of electric current can produce a wide range of conscious experiences—hypothetically, as many experiences as a brain could ever hold or have, if only we knew the code.

Of the thousands of patients who have had such surgeries or where deep-brain stimulators are implanted in various places, at least the following are possible to induce in a person with a well-placed shock to the brain:[7] whole-body swaying or rocking; blushing; urges to grasp; feelings of being "in a trance" or "out of this world"; the sense that things are not real; out-of-body experiences; laughter with or without mirth; the feeling that the body is levitating or floating; vertigo; dissociation; the sensation of "falling flat"; blurred vision; nausea; sounds seeming "distant"; blurred or distorted faces; auditory hallucinations in the form of voices or buzzing; tingling; déjà vu; an inability to count, speak, read, name things, or breathe; a fleeting, transient depression; the memory of having picnicked with a former lover; anger; fear; happiness; the urge to cry; the feeling of courage in the face of adversity; crying or sobbing with or without the feeling of sadness; stationary objects appearing to move farther away.

There is no single conscious experience that is unique and that cannot be listed, cataloged, and reproduced with electrical, physical, or magnetic stimulation of the brain. All that matters to the simulation is ease and survival, not truth. Which makes one wonder: Was Anna's surgeon mimicking the electrical program from other parts of the brain so precisely as to induce laughter and joy, specifically, or do the parts of her brain that also cause her to laugh merely mimic the surgeon's blind searches?

The Median Price of a Thrift-Store Bin of Evolutionary Hacks Russian-Dolled into a Watery, Salty Piñata We Call a Head

Consciousness is a nonproblem. I think it is one of those things that will never be answered but in fifty years' time people will look back and say, "What the bloody hell were they worried about?"

—**Sydney Brenner**[1]

We are all chimeras who have inherited the quirks of a single cell too timid to die billions of years ago. Every feeling, joy, thought, utterance, plant, and animal, extinct or alive, is downstream of that single cell's inability to give up against the odds. The brain, too, had to evolve to get here, and the stories we tell about it can be subject equally to the methods and language of history as they can to those of genetics or neuroscience.

If you look at castles built during the Crusades, you

will find holes in the walls and may, at first, believe that these were made as places to shoot arrows from.[2] They were not. They are legacies of a kind of construction no longer performed and thus hard for the modern mind to imagine. When the castle was built, there was no free-standing scaffolding, so wood logs were driven between the stones until the next layer and another platform could be added; it wasn't until the castle was finished and all the wooden scaffolding was removed that they realized there was no way, at the time, to repair the holes. Any explanation that the holes were constructed in order to shoot arrows from—despite how well suited they appear for the task, after the fact—is incorrect, and the lesson applies wholesale to castles as well as to all of biology. Many erroneous assumptions are likewise made when one looks at any modern mammalian brain and infers, hundreds of millions of years after its creation, purpose from function.

The brain evolved slowly, somewhere in the unbroken three-and-a-half-billion-year lineage from single cells until now, and is thus downstream of, and unable to course-correct from, past optimization for stressors, predators, and dangers that have not existed for billions of years. What we call neurotransmitters (the very things that pool between the brain's cells in unfathomable numbers and skew identity, thoughts, reactions, and emotions) are, simply, unused food. Dopamine is just a sautéed amino acid that could have been used for energy by a single cell but was not. Nature uses the same building blocks when it can, and what we often see in it are legacy functions from billions of years of scaffolding that we do not understand and that has, in the interim, been removed.

That cockroaches have octopamine, which is only two steps away on a molecular cascade from dopamine, does not mean they enjoy the reward from slot machines or refined coca leaves the way some humans do.[3] That single-celled protozoans and oysters and diatoms carry a gene with similar regions to the one that helps, in humans, an amino acid become dopamine does not mean that these single-celled creatures enjoy the reward from slot machines or refined coca leaves the way some humans do.[4] That lobsters have opioid receptors to bind opiates does not mean that they feel pain the way mammals do. That all studied vertebrates have opioid receptors to bind opiates does not mean that they feel pain the way lobsters do.[5]

When I once marveled over the paradox of a single molecule's involvement in both fear and sexual attraction in mice, a colleague responded, "Well, what else would do it?" His point, as I took it, was that there are only so many of life's building blocks and that each is reused—that it is the same material, the same building blocks, over and over, found throughout the web of all life. There is a sameness of being.[6] And so, to theorize about the brain's function, one must be as careful as when one is judging history.[7]

Just as a medieval castle is not built in isolation, a gene is not used in isolation. What appears to be cortical function or castle rampart or genetic destiny needs to be considered within its historical niche. For example, what we call a modern "mammal" is a stand-in, generally, for "placental mammal"—that is, an animal that gives live birth with assistance from a specialized organ, the placenta, which culls energy from the host to feed the fetus through a tube. How did we get here?

A majority of mammals, shortly after the K-T extinction of most life on Earth, were nonplacental, but in a certain select few (or maybe just one), a viral gene found its way into their genome, and future generations of mammals the world over would thus co-opt the specialized strategies of viruses and use them, at grand scale, to become, arguably, one and the same.[8] Parts of the viral gene for adhesion became a mammalian gene for placental adhesion, and this is now a permanent feature in an entire lineage of mammals, required for the exchange of nutrients across the barrier between mother and fetus.

Likewise, when a mammal is only a few cells big, a different viral gene kicks in, arranges some of the fates of each of the cells, and then turns off, seemingly never to turn on again.[9] If this gene is silenced, the cells stop growing. What does it do? Nobody quite knows. The only sensible conclusion is that thinking about a human brain in isolation from its evolutionary history and the sameness of its being with all life on Earth is almost certainly going to misunderstand its history, its function, and its repair.

Even more damning to the idea that there is a single,

human-only kind of brain or consciousness is the finding that the chemicals that slosh around in the brain and
seemingly give rise to what makes us *us* use a protein
shell derived from an ancient virus.[10] In other words, the
envelope for some of our memories is made of something that came from a virus, which is a strange and
profound chink in the argument that consciousness is
uniquely ours. But what should we do with this information? Are we human and virus or just human? Do we
count all of life's history in our product description? We
would not say the content of a letter is the envelope of
paper or ink, after all, or that an author's verses, after
publication in print, are the verses of trees. So, even
though thoughts are packaged like viruses, the thoughts
are somewhere else. But if they can be said to be ours at
all, they are ours only on loan.

Sure, an *H. sapiens* brain can, for a day, power itself
on only a cigarette and a banana and create something
that either is or is not indistinguishable from free will,
all the while keeping track of the concepts of "Russian
doll," "piñata," and "thrift store" long enough to combine
them into a thought about itself. The process through
which brains came to be, however, best resembles a
power station that had to transition uninterrupted from
coal to solar without once shutting down. Neat, sure—
but all is not perfect. The *Diagnostic and Statistical
Manual of Mental Disorders*, or *DSM*, lists hundreds of
ways that brains, like power stations, can break.

In part, the messiness is a legacy of a few moments

in early vertebrate evolution, known as "2R," when the chromosomes that contain genes, which contain strands of DNA, started duplicating.[11] A clever runaround to the problem of never being able to turn off the power plant, 2R is similar to attaching, say, a cloned power station next to the original and allowing the clone to mutate and evolve so they can keep running while trying new designs out. Likewise, what appears to be excess, or junk, in genomes (thirty-five percent of human genes are duplicates of some sort) is actually, according to the concept of 2R, a robust defense against randomness.

The brain does the same. Like a city where every large building used the same architect and blueprint, the outer surfaces of the human brain are millions of the same, repeated architectural clusters of a few neurons, over and over. Each of these cortical columns is a duplicate, which allows individual ones to specialize, evolve, and control its own small piece of the whole.[12] Brains are not the product of a single evolutionary line with any kind of plan or reason but a Frankensteinian patchwork of error, theft, malformation, crookedness, accident, and chance: like the Great Pacific Garbage Patch, consciousness goes wherever the currents take it.

The largest outstanding question in neuroscience is why activity in one of these columns gives rise to the feeling of "seeing," in another (which looks exactly the same) to "hearing," in another (which also looks exactly the same) to behaviors like laughter or kicking or crying. It gets even stranger. In one example, a girl, awake during a surgery for intractable epilepsy (the doctors hoped to take out the part of the brain where the seizures began), both laughed and experienced the feelings of merriment and joy when a part of her brain called the supplemental motor area was activated—artificially—by the surgeon's electrode. In other words, a small burst of current in one region of one column, thought to initiate actions, led to both the physical act of laughter and the joyful, subjective experience of mirth. Did one cause the other? Did the surgeon's electric wand cause both, or did one lead to the other?

How can we possibly ever expect the word "consciousness" to contain within it the collapsed variation of billions of years of evolutionary differences? The concept, like the brain that holds it, has evolved. Your inside experience is mostly similar to that of others, but it is also surprisingly variable both within and between people in a way that language rarely captures. People can go their entire lives without realizing they or their close partners are synesthetic, with some senses mixed such that letters are colored just because they grew up with Fisher-Price alphabet sets, or aphantasic, with no mind's eye at all.[13] James Joyce said his mind was like a grocer's assistant; Einstein that he could imagine chasing a beam of light. Some people, when they close their eyes, see nothing. Some can practice chess in their heads. Some hear words as music. Some have perfect pitch. Some have tinnitus. Some think consciousness flows like a river. Others, that it sits stiller than a tree. There is no one such thing as "consciousness," and the attempt to study it as a singular phenomenon will go nowhere.

In physics, things generally have to be on time, especially if they are causal. In biology, there are small delays as the molecular gears start and stop and as the chemicals slosh around. If the sound coming from a movie theater's speakers is mismatched by less than one hundred milliseconds from the visual action, it will go unnoticed by all persons watching because the brain is stitching together all its selves all the time, washing over the biological delays to create a binned, single self.

Likely, there will never be a Galileo or Newton that comes along and solves dementia, consciousness, or sadness in a single set of equations. The brain exists as a series of learning rules and is guided, like the square watermelons of Japan, to fit nicely into the box it is given. There will never be a scientific model that explains all of biology or all of the brain or the way in which neurons generate consciousness or are themselves conscious. Every single neuron, which grew into its own patch of brain, is itself a moving creature, with dendritic fingers and moving organs with more misexplained castle holes in its ramparts than stars in the universe. The secrets of the brain will be a tapestry of scientific truths, not a theory of everything.

Sunlight Raining Down
on Gridworld

Raised by wolves, a brain learns wolf culture. Raised in the Ice Age, a brain learns Ice Age logic. Future generations of human children born upside down in space will figure out those rules, too. A lot can be learned by learning.

Neither intelligence nor consciousness are required for learning, however. Imagine the different ways a thermostat might regulate a room's temperature and learn facts about the world in the process.[1] A simple way for a thermometer to start off would be to compare current temperature to a desired temperature and make a few rules:

1. If $T_{current}$ is less than $T_{desired}$ → increase heat.
2. If $T_{current}$ is greater than $T_{desired}$ → decrease heat.

Eventually, inefficiently, something like the correct temperature is settled on. But a smarter thermostat, given additional control over temperature-relevant objects in the room, could do a much better job by predicting what might happen when it acted.[2] At first, its predictions would be based on simple observations, like the room gets colder when a window is opened or during one half of the day in twenty-four hour cycles (night/day). Over time, and over days and years, as the observations become more complex, so, too, will the predictions become more complex. When the window is open during part of one of what appears to be a yearly cycle, the room gets either hotter *or* colder (summer/winter). Eventually, the thermostat, knowing absolutely nothing about the features of the outside world and knowing nothing about wind, sun, gravity, or Isaac Newton, could learn things about the seasons, about the differences between night and day and cold wind or hot wind, and about the local effect of moisture on glass. Most important, it could start to make guesses about the outside world and test them to learn what happens at the edges of the seasons or about the metabolism of a roosting bird accidentally let in through the window.

In 2009, a computer program, Eureqa, using machine-learning algorithms, was able to derive a close approximation of one of Newton's laws of motion using a similar strategy.[3] With only a robotic arm to push a double pendulum (a pendulum with a second pendulum hanging off it), it was able to find order within the highly chaotic movements of the pendulums, which showed that something as invisible to common experience as the physical laws of all motion, which apply equally here and on the moon, can be learned by rote observation, prediction, and action alone.

Of course, brains already knew this.

All brains, and every nervous system, are in a similar predicament to the smart thermostat stuck in a room. A brain only knows what happens to it. Nothing more. All of its conclusions are drawn from the same kind of logic that the smart thermometer in the room used—all that can be known is the impact on neurons (temperature, in the analogy) based on one of its movements (say, opening the window, in the analogy) compared with its assumptions about what might happen based on everything learned so far.

Nothing more about the rules of the world is available to any forms of life on Earth, all of whom must continually consume anti-entropy, or order, to keep their wheels spinning.[4] We cannot eat rocks, or sand, for nourishment, even though we are all just made of carbon, because we the living must eat only those fellow combatants in the Four Billion Years' War. When we eat a plant, or some yeast and flour that we alchemically convert into food, or any meat, we are really consuming armament in the unbroken chain of the cellular fight against disorder. We get no nourishment from rocks because they, as yet, have put up no useful struggle.

And because all we can do is flee or follow other life, our cognition has coevolved with the patterns of how life appears, moves, and disappears on the outside world's grid.[5] We know how to search because the objects of our search are not uniformly distributed but are patchy, legacy GPS coordinates of where the sun once was. So, too, we think because we need a way to efficiently search and are conscious of some percentage of this thinking, all so that our unique pocket of order can find, flee, or feast on other pockets of order. Death reverses the small victories, a return to nonliving, nonfighting dust. We know this because almost immediately after a human heart stops, algor mortis—the "death chill"—sets in and the body begins to cool down 1.5 degrees Fahrenheit per hour until it reaches room temperature.[6]

The entropic terms of the fight are thus set for mammals of our approximate size and metabolic need. That is, we do all that we do in order to prevent 1.5 degrees of decay per hour, every hour, for our entire lives. Nothing more. There are many strategies in this fight, which began when a glom of something made a wall around itself and created a gradient of inside-hydrogen versus outside-hydrogen, which is to say a gradient of pH, which is to say a gradient like a spring that, when released, can power a small pump with the power of a lightning bolt, as if a molecule-sized dam on a river of protons.

The brain, which is composed of hundreds of billions of these borders, each of which we call a cell, and which contains many hundreds of trillions of these dams, which we call proton pumps, takes all of this stored-up potential and turns it into a learning device whose purpose is simply to help learn some things about the world despite the fact that there is no prediction of the universe that does not end with a loss.

Until then, we do what we can. Like butterflies that migrate over multiple generations to a single tree in Mexico, most of whom were born on the migration path and die before ever seeing its end, we wake and behave with approach-avoidance drives in an innate direction we, too, may never know. To do so, the brain tries to predict, based on its experience, what will happen next, which means that consciousness is neither an accident nor an epiphenomenon. It is one of the greatest tools available in the hunt for patchy spots of anti-entropy.

But what about those predictions that happen too fast for conscious awareness? Consider the following story, told to me by an executive in the U.S. intelligence community, about a time in the late aughts when he was on a dangerous road, in Iraq, being driven by a soldier:[7]

> I was on the Baghdad airport highway to the Green Zone, and we're driving along, and my driver, this enlisted man, this kid, slams on the brakes, cuts across the median, and heads back to Camp Victory. And I said "What happened?" And he said "I don't know, something felt wrong." He was very upset. We-were-going-to-die upset. I let him go for a while, and later I said, "What do you think it was?"
>
> He says, "There were no kids. We drive that

same route every day at the same time and there are kids kicking around an old soccer ball, in that field, and today there were none. And that felt really dangerous to me. And thinking about it, it's because the moms know when the bad guys have planted a roadside bomb, and they keep their kids away."

Even on the simple point of whether this story is about consciousness at all, there is disagreement. The nays might say, "It is a story mostly about unconscious processes." The yeas might say, "If so, how did the young soldier's intuition get trained? How was he able to access it retrospectively, hours later, if it was not conscious? Is there some sort of postal system whereby unconscious motivation is sent to a different, conscious part of the brain?" The nays might respond, "The soldier's brain was just making up the reason, anyway, by confabulating a plausible story."

Or consider Anna, the teenager who, during brain surgery, made the equivalent of a rapid cognitive U-turn. While awake during surgery, with her skull opened and a neurosurgeon behind her, probing its topography for clues, Anna noticed suddenly that something was not right. She laughed. Or she found herself laughing. But why? In the surgical room that day, she had laughed without a reason, and seconds later, when asked why, she gave an answer that plausibly explained the anomaly she witnessed. Just as the soldier plausibly explained why he made the U-turn.

In their actions, both Anna and the soldier exploited a small piece of brain hanging off the back-bottom parts—a "littler brain"—which contains in it more than two-thirds of all their central neurons and which has the capacity, though it remains unclear how, to keep track of everything that has ever happened to them.[8] This littler brain is billions and billions of repeated motifs, many with a one-way connection back to the other parts of the brain. Anytime one learns a new skill under conscious awareness—to ski, skate, play piano, say—that awareness fades into the background of the littler brain with practice. This is intuition, not magic, and over time we have all also become experts at intuiting our own reasons for doing things, no matter how far from the truth the answer is.

Intuition has a maligned reputation as one of the lesser kinds of reasoning but is, in fact, second only to consciousness itself as the mammalian brain's greatest feat. Intuition is the reasoned product of a lifetime of careful, metabolically expensive observation. It is the output of the brain, never the gut. (Do not trust your "gut," ever, unless you need to dissolve something quickly.)

Intuition works so well because the brain is predicting everything all the time and the ringing red bell of alarm that the soldier experienced on the road in Iraq and that Anna experienced when she laughed for "no" reason was a mismatch in their observed statistics of how things should be. Nearly *two-thirds* of the brain's neurons are devoted to prediction and feedback so that the brain can learn and update the validity of previous predictions.

The whole reason for my and your and Anna's and the soldier's bodies to be around ninety-eight degrees Fahrenheit is to insulate the thermometer and its room; to keep it warm so that fungus doesn't grow as it does on amphibians and so that our warm-blooded enzymes, and thus thoughts, can work optimally.[9] When a warm-blooded animal gets cold, it shivers its muscles rapidly, a process that turns stored energy into heat more quickly than at the normal rate. There is little difference between shivering and thinking, metabolically—both are muscular mitigation tactics in the effort to learn just a little bit of useful data about the very universe that tries to pull us apart, every second, every day, until the windows stay open and there is no more difference between inside and outside reasons.

An Ante Meridiem Radio Drama

A horse is a horse, of course, of course.

—Mr. Ed

Movies often get insect vision wrong.[1] Having eighteen eyes makes eighteen points of view no more than having two eyes makes two points of view. Each of our two eyes takes light from slightly different but mostly overlapping sections of the world in front of us, and these combine into a single cyclopean aperture.[2] If you cordon off each eye by placing a divider between them, however, and show each eye a different image, the illusion that consciousness is concerned with only a single, unified point of view will quickly be broken.

People who have experienced this eye-line division in a laboratory setting report viewing, at some cyclic rate, usually every few seconds, an alternation back and forth between the two images.[3] If you show the left eye, say, an image of a cat's face and the right eye an image of a small house, many people would report seeing a cat for a few seconds and then a house and then back to the cat and back to the house, ad infinitum. Only one eye seems to get priority, each at a time, as if siblings competing for the top bunk. One eye's input rises to the level of consciousness as the other's gets shunted into the background.

This is not how it has to be. It is possible to imagine other ways the brain might combine two different inputs, one to each eye: the rival images could overlap and coexist, like doubly exposed negatives; they could alternate as background and foreground; they could blend into each other as an average of color, luminescence, or perceived motion; they could combine as the average of the distance between the centers of their prototypical categories in semantic space; they could combine into an object halfway, alphabetically, between "cat" and "house" in the person's native language; they could combine so that the facade of the house resembled a cat's face with whiskers, windows, roofing, and ears, and all to scale; they could be bright or dark proportional to positive or negative personal history with cats or houses; they could merge into a visual portmanteau, a centaur-like cat-house; they could be separated by the parts of the brain that keep track of space so that they exist far apart rather than centrally or atop each other; they could both coexist but shift in and out of focus with attention; they could disappear, shoved into a blind spot of incongruity or impossibility; they could rotate as a sphere, as if on a turnstile projecting out from inside a snow globe;

the bigger one could win; they could make a sound every time they switched; each could have a leitmotif, as in a Wagner opera; because the brain is doing its best at all times to tell the simplest possible story about the world it knows, and because two physical objects cannot coexist in the same place at the same time, they could displace the less obviously massive object to the side or underneath the more massive one, thus placing the cat a little to the side of the house; they could melt into an object that is only the parts of the cat and the house and that, when overlaid, are exactly identical, as if a kind of perceive-by-numbers kit.

In fact, depending on the features of the images shown (and perhaps the person they are being shown to), conscious perceptions are occasionally incomplete (the "ignored" image remains slightly visible), piecemeal (patches of both appear at the same time and interwoven, like a two-color quilt or a completed puzzle from two different sets whose pieces happen to fit), or strangely contrasting (the "ignored" image does not disappear but is merely low contrast).[4] Why is this how such rivalries are settled, sometimes Solomonic and sometimes not? Is there something about consciousness that requires the two images to flip back and forth, as if curtains are being rapidly drawn and closed on multiple, co-occurring stage plays?

Imagine if a targeted mutation or the whims of evolution caused the growth of eight instead of two eyes on a mammal, stacked vertically, into two columns of four, like domino pips. The same image-flipping experiment could be performed but with a different image for each eye. Would consciousness flip back and forth between all eight or only a select few eyes? Would the switches happen at some cadence determined by physiology and statistics of the brain or by chance, or from the fatigue of the neurons holding it all up? For the sake of argument, let us say the person reports consciously "seeing" each image flipping in the following order, corresponding to each of the eight eyes: 1, 4, 4, 2, 4, 5, 8, 7, 6, 1, 7, 1, 2, 3.

Any good theory of consciousness should be able to explain why 4, and not any other eye, followed after 1. Each eye had competed for the display rights the same way that any object wishing to leave Earth's atmosphere must—that is, by achieving escape velocity and overcoming the physical forces that prevented it from doing so. The visual signals from the winning eye were made available to large, important swaths of brain during their return trip from the front of the brain to the back and, just as a radio antenna's broadcast range is determined by the kind of message it wants to send and its size, so, too, does a pixel of consciousness arrive via broadcast from the antennae that reach the farthest.[5]

One of the more confounding results from a mid-1990s study was that a teenaged patient, looking at a picture of a horse while an electrode stimulated her brain, laughed, responding that she had done so because "the horse is funny." In fact, the horse was not a funny or funnily drawn horse. In fact, the joy and mirth and laughter she felt was a coincidence. However, because there can be only one show on at any given time, the patient responded as she did because both the visual representation of the horse and her stimulation-induced laughter were globally broadcast, like a radio or television signal, into her consciousness at around the same time.[6]

If she had been shown a picture of a cat ten milliseconds before being shown the horse, the part of her that can self-report things would not be able to report that there was ever an image of a cat because broadcasting has rules, a kind of FCC controlling the airwaves.[7] And, like any television or radio signal broadcast into the air, conscious impressions must be both sent and received. Ultimately, this broadcast is useful because it acts as a kind of buffer that allows movement to be coordinated and planned so that life can find, say, water beyond the horizon and out of sight. It is here also where all conscious thoughts happen, to allow planning for things. The horizons may have shifted inward as the signals bounce around the brain like echoes, but the broadcast technology has stayed the same.

The purpose of consciousness is to make information broadly available to wide swaths of the brain that need it. The act of projection is not just the onstage mise-en-scène but is consciousness itself.

A Small Town with Too Much Food

Once upon a time, four billion years ago, there was one source of heat, a large fire, during an otherwise cold winter.[1] It was hotter closer to the fire and less hot farther away. Interesting things happened at this line between hot and cold. Nothing stayed alive without it. In order to be farther away from the fire and away from the gathering crowd, it was necessary to somehow bring the hot/cold border along. Some built single-story huts to hold on to the hot/cold border created, at first, by the fire and winter. If someone had walked inside the walls of one of these one-story huts out far from the fire, the walls just faces of repeated brick, there would be nothing much inside. The inside of the hut did nothing but cling desperately to the fact that it was not the outside. The hut passively took package deliveries, small spiral statues that it shoved into its walls, breaking the statues apart and combining their parts all just to stay upright. Hundreds of kinds of small spiral statues passed by the hut outside, but most were ignored. Those that were not ignored arrived passively at the doorstep only infrequently and by chance. Soon there were small nets placed out of the hut's windows built from statue pieces, functioning a bit like spiderwebs, to catch more statues in its grip and reel them in. A small nearby hut offered a deal. It was sick of moving around and agreed to handle all the package deliveries and statue piece-building wear and tear as long as it could set up camp inside your hut. It promised not to take up much space and it kept its word, setting up in the corner, in its new home. Now your hut is unencumbered by box openings and free to focus on finding more spiral statues. It has been way too long at this point to try anything else. It was nice by the big town fire, you recall, but independence has its own perks, too. There were many other huts all of a sudden. Some huts clumped together near each other, stronger in numbers. One day, the hut you are in receives a strange package. It is *part* of a spiral statue, like the Maltese falcon with its head torn off. You assume it was sent on purpose from a different hut and just happened to land on your hut's door. Maybe they were trying to tell you something? You decide to keep it. The occasional spiral statues still arrive by chance and still get shoved into the corner to do whatever fancy metallurgy the small corner hut does, all to keep you from bursting. You notice one of the spiderweb-like nets jutting out the window is now wrapped in something stronger than ever before. These start pushing into the ground, like oars. The hut is moving. You are moving. Your hut moves toward the other huts that keep sending you the statues with their heads missing. What are those huts trying to say? That they have so many spiral statues nearby, so they can afford to waste some? Might as well find out. What do they want in return? It is true! Some huts had found so many spiral statues they could afford to cut some in half and send a floating amalgamated message, the logic of largesse: Here Be Spiral Statues. Other huts like yours, now also with net-oars, lumber over after receiving the same message. Your hut moves itself into an odd but sensible position with its door facing the spiral statues and a small outgoing-mail slot on the other side, where it sends its own fused statues made from melted statue parts if and when it makes them. For every one fused statue it sees, your hut sends out sometimes dozens just like it. You hope the other huts all get the right message, the only one your hut knows: Here Be Spiral Statues. Gah! Predators. Huts are being torn apart, prodded and poked by other huts. You see carcasses everywhere. Debris fields full of spiral statues and even larger, twisted and fused statues you've never seen before, but still all made from parts of the original smaller ones. In order to survive predators, the huts start huddling tighter, like water or starlings, leaving predation to chance. But not entirely to chance, because a few huts decide to connect, grab on to each other's net-oars, and act as a team, like a town. Light-sensitive rewrapped net-oar huts collect on one side of town and connect with a new, superlong rewrapped net-oar to huts on the other side of town. A single entrance to each multistory building. You go in one. You notice stairs in the lobby of one, a handwritten sign that says: VISION. You go to the second floor and see light-sensitive rewrapped net-oar huts everywhere. So that's where they've been hiding. You leave the building and look around. The townwide dome gets much, much darker. The dome huts, in fact, block almost everything. The light-sensitive rewrapped net-oar huts, desperate, move to the tops of the buildings and try to poke through, toward the light. You turn back and the scattered buildings are gone. Instead, you see one huge building, six stories high, in the center of town. You go in and find five elevators in the lobby. The whole building is made of millions of superlong rewrapped net-oar huts. Each of the elevators has a button: VISION, HEARING, SMELL, TASTE, and TOUCH. You go to the VISION floor, as before. You recognize small huts covered in spring-loaded wind chime traps. The sound is deafening. Fused statues are being sent back and forth at such a rapid speed you can barely step without tripping. Many huts simply pass the statues back and forth between them, over and over and over, hundreds of times a second. These huts connect to the light-sensitive rewrapped net-oar ones near the top of the dome and to stretch-sensitive rewrapped net-oar huts nearby the net-oar ones all over town. None of the floors of the building connect to each other, so you quickly get bored. It's the same six-story building but now it is different. There are five elevators in the lobby, each with its own button—VISION, HEARING, SMELL,

TASTE, and TOUCH—that goes to its own floor. While exploring this time, though, you realize that a few of the floors connect now via stairwells in the back. For example, VISION connects to HEARING; SMELL, to TASTE. Entire floors seem to be dedicated to making sure the superlong rewrapped net-oars cross nicely and that their message gets sent. The building in the center of town is now one thousand stories high. The huts have become individual rooms, separated with the same repeated brick as always. The same brick that was found in the first hut, so many billions of years ago. The elevators in the lobby each have a familiar button: VISION, HEARING, SMELL, TASTE, and TOUCH. You discover now, to your delight, that each elevator can go to most of the other five labeled floors. They seem to cross in interesting ways. The other one thousand minus six floors are all connected in bizarre and intricate ways, too. All still made out of superlong rewrapped net-oars, but each hut now has hundreds or sometimes thousands sticking out. You get lost trying to follow one superlong rewrapped net-oar all the way through the building since it seems to cross through many floors and pass by many other huts. You notice a few special floors along the way, unconnected to the elevators, which seem to connect to at least 80 percent of the other floors. Many of these also just send statues of all kinds back and forth to each other hundreds of times a second. There are lots of connections between those rooms now, too. In fact, most of the one thousand floors are not even connected to any one of the floors and realize that the messages coming in are enormously complicated. How could anything possibly make sense of them all? You notice spherical breaches in the dome above the building at the center, which are full now with light-sensitive rewrapped net-oar huts, are coordinated with the net-oar huts. You climb all the way and make it to the breach in the top of the dome. Every time the spherical breaches full of light-sensitive rewrapped net-oar huts stop moving, which is at least three times a second, you can see. Some of this is preplanned in other parts of the big building in the center of town, and the same part that sends the message using exquisite combinations of statues ("Net-oar huts, do whatever you do to make the spherical breaches move to the left") also sends a copy to other floors ("VISION floor of main building, please be warned, I just sent a command to move both spherical breaches a little to the left") so that it can plan for the perceptual shift. Thus, the spherical breaches are showing the outside world as it is, but only through rapid flits and very minor but near-constant movements of the entire town do the objects that make up the world appear to be stable and the inner hut, and thus you, Anna, are the agent of fluid motion through them until one day the power keeps going out unexpectedly in some parts of town and, years later, the entire dome covering the town is lifted and, hovering above it, a giant new dome with another small town behind it that has too much food sends a platinum steel pole down into the building in the center of your

The Arbiter of Elegance

Elegance is refusal.

—**Coco Chanel**

At all times, a brain is taking in more information than it can handle, with more possible ways of configuring itself than the universe has atoms, which means every brain needs a way to keep track of itself in order to keep things tidy and efficient.

In *The Satyricon,* Petronius, at the time Nero's official arbiter elegantiarum, or "arbiter of elegance,"[1] wrote against the rhetorical excess that had infected Roman youth:

Action and language, it's all the same: great sticky honeyballs of phrases, every sentence looking as though it had been plopped and rolled in poppy seed and sesame.[2]

Petronius's lament, as I read it, had to do with excess in both action and language. He is right about at least one thing: language and action are, indeed, the same as far as much of the brain is concerned. Both are gestures put on by the brain's only mechanism for interacting outside of itself, muscles, and excess in either is both aesthetic and metabolic waste.

But how does each kind of life on this planet choose what is best for its needs? Primates are exquisite movers, somewhere between a plant (stationary), a sea slug (semimobile, but only for a time), and a bat (very mobile) on the total scale of ways to move at any given time, a concept called "degrees of freedom." Because plants are so limited—with relatively few degrees of freedom for movement—and have no nervous systems, they tend to have large genomes to compensate.[3]

In part, this is because plants cannot move in the face of danger and so instead of a complicated, expensive behavioral menu of options for finding rather than being food, plants have a complicated, expensive genetic menu. Literally rooted in the ground, they still want to do their best for defense against predation or random injury, and so, in the face of a new danger, many plants can scramble their genetic, and thus chemical, makeup in the hope that something works quickly enough for either them or their neighbors.

But not every creature was content standing still. Neurons, which are tuned either to orient toward food or flee before becoming it, changed the game. The sea squirt is an oceanic creature the size of a toddler's boot with a few thousand neurons, which lands once and sticks on that spot the rest of its life.[4] It then dissolves its own nervous system so that it may grow a new, different one more useful for its new life of immobility.[5]

And on the farthest end of the mobility scale are bats, the only mammals who fly and echolocate, two feats that combine into an extraordinarily complicated effort. Bats end the day so taxed—flight alone can raise their temperatures to 105 degrees Fahrenheit—that the insides of their cells literally break down during rest, spewing DNA that the immune system then has to clear.[6]

Each of these creatures—the plant, sea squirt, primate, and bat—has a different set of possible movements to coordinate. The multiple possible plans are not wasted metabolism for unused ends because nothing about life is fixed anywhere along the line. If the outside world's events were already laid out and determined, all such planning would be a waste. A menu of preheated options for action is better than starting from scratch.

We thus inhabit a world of expert make-believe, constantly planning and playing in counterfactual scenes of projected movement, reward, and action. Some parts of the brain are constantly planning dozens to hundreds of movements in case any one of them should be imminently useful; only some are chosen, and the unnecessary ones are, eventually, taken away. Neurons or groups of neurons that control these muscular actions are sensitized to act, like a preheated oven, all so that those particular neurons of a certain action can become active or integrate incoming information quicker.

Our assessments of what to plan, and when and why to plan it, are reduced from an outside world that is actually quite regular in its statistics. Objects tend to stay put. The sky tends to stay up. Farther-away things tend to appear bluer. No animal moves faster than we can see. Our possible actions are thus based on prior experience with similar sights, sounds, and locations learned since birth, with strong curation by the evolutionary set of sensing tools and the physical limits of our bodies.

For most mammals, given the near-infinite options afforded by movement, an elegant, dimensionality-reducing solution is required.[7] And so the brain reduces its expenses to only the variables it cares about and in the process simplifies both itself and its view of the outside world. For example, though there is a near-infinite number of possible colors, the human eye receives information about only three color dimensions—short (blue), medium (green), and large (red)—based on what the cells in the eye respond to.

There are many other clever solutions, too. If we think of every neuron leading to a muscle as a puppet string, each string can only be tugged on once every few milliseconds, at most, which greatly reduces the coordination problem. As well, muscles often act in coordinated patterns, which even further reduces all the possible paths.

Consider Anna's laughter and stated reason for it, when asked: that she experienced joy and mirth. Though only a tiny fraction of her brain was stimulated, it started a cascade of coordinated downstream activity. When Anna laughed, and she was asked *why* she laughed, she did not say: "Well, I laughed because the neurosurgeon used an electrode to stimulate the part of my brain that caused a group of neurons to coordinate their firing and output so that my throat bounced air around in a stereotyped and repeatable pattern." Instead, she said she laughed because "you guys are just so funny . . . standing around" and also because, while viewing a picture of a horse, "the horse is funny."

We know these are not correct answers. They are, however, efficient answers.

The story Anna's brain chose was tacked on afterward with little regard for the underlying, discarded possibilities. It is highly unlikely, in most evolutionary or life circumstances, that a neurosurgeon is pulling the puppet strings of your laughter, joy, and mirth. So why did she choose these answers? Notice that they are not random. There is a kind of logic to them. They include features of the world that may at one time have been, or could be, funny. Horses *can* be funny. People standing can be, too. Anna did not say the colonoscopy camera hanging on the wall was funny; she did not say she laughed because she remembered a joke. Likely, her brain chose reasons among likely options, not an infinity of options. There is a structure to our reasoning based on experience. If you ask someone to imagine a clearing in a forest and to place in the clearing in their mind's eye a wooden log, a key, and a bear, the bear will likely have moved while the key and log will likely not have moved. In other words, some objects (bears) are more often presumed, because of real-world experience or inference, to be moving and some objects (horses, people) are more often presumed, because of real-world experience, to be funnier than others.

As well, anything within Anna's arm's reach gains outsize importance as the visual or acoustic chance of a predator is calculated by some parts of the brain, which are being fed information by other parts on the likely depth and texture of the walls and objects and their probability for movement. The inevitable selection among these possible choices is not historically set by Anna's prior choices but is sieved by a mechanism in her brain that is the sum total of the interactions between the preplanning and the outside world's local options. The number of these interactions crosses, at some point, a threshold, and the outcomes from this process are what feel like free will, but the path to behavior was as elegant as Petronian prose. Every action is not by fiat but rather the remainder when all else is removed.

Neurons are just cells like all the others and are always "on," unless damaged. Anna is using 100 percent of her brain, always, even during surgery, and though most neurons do not know it, they exist only to produce or simulate action. As she learned and became better skilled at any or all action, her brain must have somehow, through all its sensory streams, and in conjunction with the six-hundred-and-fifty-dimensional muscle-coordination problem, figured out how to do the same thing more efficiently next time.

Movement and thought are braided together throughout all life on Earth. When a brain thinks, it is acting on itself. A neuron does not know or care whether there is a muscle or another neuron on either side of its synapses. It discharges its neurotransmitters or action potentials all the same.

Right before Anna laughed, the electrode stuck into her brain had short-circuited the normally occurring path from plan to action. Her brain was not used to actions without plans. When Anna thought of a reason why she laughed, her brain had to plan possible answers before settling on the real one; when Anna spoke this choice of reason aloud, her brain had started by initiating a movement from one of the plans from its parts that control the muscles that control speech. Her imagined speech, or "inner" speech—the voice in her head—is just a kind of planned movement with no muscular output.[8] Thoughts of the kind that are imagined speech, either the production of or the reception to, or that involve conversation, are therefore also simulated motor acts. The general recipe for these kinds of thoughts—take an action and then silence it before acting out—works for all the thoughts we call thinking.

Anna's brain was comparing its expectation to reality multiple, perhaps thousands, of times a second and, in the absence of certainty, her brain guessed based on likelihoods from prior experience. The fullness of the reductions required for her brain to arrive at its wrong answer was an elegant story, with the unnecessary things taken away, pieced together from disparate kinds of information.

It is of no use for a brain to conclude that there is no reason behind any given action because the experience would be uncategorizable for later use or comparison. As the brain continually tells itself stories based on incomplete data, it becomes an expert at the heuristics of compression; as these become habitual, the in-brain comparisons required to achieve them get off-loaded more and more into the unconscious parts of the brain, like the little brain hanging off the back-bottom parts. The more these comparisons get shoved into the rapid unconscious, the quicker they are and the more elegant they become, hiding as intuition. The parts of the brain that keep track of movements map efficient lines, like expert sailors through a storm, through planned and acted-out movements, and this same part of the brain is co-opted to keep track of ideas, concepts, and memories, meaning that place-based recall of memories is no different from place-based recall of ideas. The effort and energy required to move between idea A and idea

B is, in the mind's eye, the same effort and energy required to move between place X and place Y.

Like consciousness, elegance has no one true meaning.[9] It is, instead, what is left over when all else is removed.

Swinging Through Ancient Trees While Standing Still and Hearing Voices

If a lion could speak, we could not understand him.

—Ludwig Wittgenstein

Just as there could be no job listing for "train conductor" before the invention of trains, there was no speech, and thus no speaker, before the human brain started talking to itself. At a certain level of practice, and with a few new wires added onto the basic primate growth blueprint, the internal commands that once created hand gestures also created language.[1]

Those extra few wires, which connected a chunk of brain required to swing through trees to the laryngeal cells that control muscles in the throat, meant that the muscles that allowed air to be bounced around and leave the mouth in various states of compression were not coordinated by dumb instinct but could learn a level of expert articulation as easily as the hand could. Thus, it became to the brain, whose only recourse is to move well, that swinging through an actual tree was treated no different from swinging through a linguistic tree except that the latter, unlike the former, is infinitely branched.

Only a small handful of mammals and birds—such as humans, dolphins, bats, elephants, seals, parrots, songbirds, and hummingbirds—show the capacity for vocal learning, which appears to be required for both complicated language and the ability to coordinate whole-body movements to sounds, as in dancing.[2] Possibly, in *H. sapiens,* the learning loop responsible for training the hand duplicated one too many times, bringing with it a small reroute, like getting off the interstate and doubling back an exit, and in so doing allowed the echoes of sounds from without and the sounds from within to bounce around for a few more seconds than normal. It is no coincidence that vocal-learning regions in the brains of all vocal learners are either within or adjacent to movement learning, which we can infer because, in patients like Anna who are awake during brain surgery, stimulation of certain language areas will stop both any speech and hand movements instantly.[3] What we do not know, but can guess, is that her thinking itself slowed, too, during this time.[4]

Brain size does not matter for language ability. Hummingbirds can imitate complex sounds; chimps cannot. Songbirds dream of singing; chimps only of songbirds singing. Likely, the different wiring from the eyes and the ears through the brain changes what is and can be done with the signals. There is no binaural rivalry for hearing, for example, exactly like binocular rivalry for vision, and it seems important somehow that speech sounds change four or five times per second and that the eye moves about to glance at its world at a rate of three times per second. Somewhere around this range, a couple times per second, which is between a little faster and twice as fast as it takes for a professional baseball pitch to reach home plate, is the cadence of information our brains like to receive and learn from.

Right now, for example, you are likely hearing many voices. You are almost certainly hearing a voice from this page as you read along. If you started speaking alongside this voice, you would also hear your own voice, which would sound as if it were coming not from outside but from inside. You also hear a running track of what these words mean to you and their associations built over your lifetime of association. I can stop talking, and if you wanted, you could replay this sentence perfectly in your head for a few . . .

. . . seconds afterward. Sounds can thus bounce around and a few seconds of tape can be written or read, and meanwhile the tape contains in it what might be said next, what could be said next, what will be said next, and also what we might expect to hear others say next given what they've already said. It is a grand game of Mad Libs. These moments in evolution when the new tape deck allowed sound-based learning, when gestures became thoughts, brought with it a Cambrian explosion of thoughts that, like mammals, took over most of the newly available niches.

We are overridden by voices in our head but are rarely tortured by them because they feel as if they come from the safety of indoors. These voices can echo thoughts, be thoughts, give commands, provide running commentary, be negative or positive. We all have them. We can listen in order to make more of them. From a small sampling of the world, a one-dimensional beating of the eardrum, which only has two binary configurations—in or out, on or off—we can extract a signal in electrical terms, turning them into memories of old ideas or even memories of new ones.

But on the inside, while hearing nothing, while the eardrum remains silent and still, we can also somehow imagine (hear) people say our name as they walk by; we can imagine (hear) them talk from the backseat while driving alone; we can imagine (hear) conversations with imaginary others—sometimes dead, sometimes fictional—during the menial morning routines of showering or drinking coffee; we can imagine (hear) critical voices, like the voice of the parent in your head that demands we use a coaster or separate, in laundry, whites from colors; we can imagine (hear) their commands; we can imagine (hear) them consider, on balance, right from wrong; we can imagine (hear) reasoning point and counterpoint.

In a study asking questions of three hundred people who had recently lost a spouse, thirteen percent said they often heard the voice of their spouse.[5] These and related phenomena are all downstream of those moments in evolution when parts of the primate brain most responsive to other- and self-generated primate hand gestures copied themselves and, in so doing, became responsive to and learned how to make acoustic gestures.

But there was a problem with the new design. Suddenly, an *H. sapiens* brain could break in a way it could not before. While swinging through the trees, missing a limb by a few milliseconds meant injury, inconvenience, or death, depending on the height. But what about a mistimed thought? As there were no train accidents before there were train conductors, there were no bouts of schizophrenia before there was the ability for the brain's acoustic loops to derail.

Psychosis related to hearing voices is difficult to diagnose even today, and impossible to do so historically, but observations of some of those who have schizophrenia give potential clues. Schizophrenics can hear auditory hallucinations of others in positions of high power, like the devil or, in one case, Richard Nixon; they can tickle themselves; they are often very, or overly, literal when asked to explain metaphors, phrases, or sayings like "Loose lips sink ships"; after receiving a warning tone, they startle more easily to a second, louder one. The number one nongenetic risk factor for schizophrenia is migrancy; many of the other risks involve being born in or living in urban environments, and there is a Tolstoy effect,[6] where moving to a rural area can remove the increase in risk. There has never been a documented case of schizophrenia in someone blind from birth; Albert Einstein, James Joyce, and Bertrand Russell all had children diagnosed with schizophrenia.[7]

The connection between thought and movement is not a new idea. There was a brief era, in the mid-nineteenth century, when doctors at Hôpital de Cery, in Lausanne, Switzerland, would diagnose people based on the art they drew, which makes some sense only if disorders of thought are disorders of movement on the landscape of thought. Perhaps, the logic goes, a pattern in brushstroke, like a pattern in speech, can be askew. Poets, for example, who use first-person words like "I" in their published works are more likely to later kill themselves,[8] and examples of phrases from those with schizophrenia are, perhaps, just mistimed grips onto the branchings of language: "rectangularly speaking," "the mineral of its substance," "an x-axis in origin," "posterior pronunciations," "the outform of the map," and "a nonverbal misrepresentation leading to an unformulated thought."[9]

Is there anything in common across these utterances, as there might be in the visual art of those with schizophrenia? Possibly. For hundreds of years, people studied electricity before finally figuring out what it really was.[10] Early on, some had thought it a liquid and tried to bottle it, while data showing it was related to magnetism were ignored for lack of an explanation.

Only much later, with a full theory in place, did it make sense to go back and study the results of the dozens of electricity-related phenomena that were not well understood or thought to be unrelated. One day, the same will happen across the many ways the brain can break.

Only parrots and humans are extraordinary, even among other vocal-learning species, in that they may have an even more complex, extra motor-learning loop that both precedes and allows chirps within chirps and language within language, but the added complexity of these interactions means that the brains had to become ever more careful about staying in tune and that there is no weakness of constitution or self during a psychotic episode but, perhaps instead, a small mistiming in the orbit of electrical currents that give rise to, or are, thinking. When Mercury's orbit was slightly off by forty-two arc seconds every century at the point closest to the Sun, many explanations were given for how and why until Albert Einstein came along with the idea that space-time was curved just a little bit by the Sun, which could warp the area around it. Mercury's timing was never "off"; its orbit simply was what it was and we just didn't know enough, at the time, to know what we were looking at.

Likely, any mistimings have a physical cause. That brains can break is a given; that when they do, they do so in ways similar to other brains that have broken is not. Consider, for example, the spaceman helmets from older

Lego sets, which would almost always break right where the plastic molding connected in the front of the helmet, near the chin strap, a bit near where the Lego figurine's jaw would be. Almost every kind of broken helmet was broken at exactly that spot, which is data for the inference that there was a weakness somewhere in how the helmet was made and, most important, for the inference where it might break again. Might a brain, equally, break the same way as any physical object, at its weak points? A common kind of stroke—middle cerebral artery occlusion, or MCAO—happens in the same spot over and over inside people's heads, right at a major arterial fork, where the blood flow is forceful and splits off. Through something like erosion, no different from an ocean's effect on a coastline, the area weakens. Some people, defined by their genetics, have different outlet flows at this area in their vasculature, which can protect against MCAO stroke; others, also defined by their genetics, do not, and they are at risk of their brains breaking at just this spot. The brain is no more or less physical than any other machined toy: a headache is not the absence of ibuprofen; a broken space helmet is not the absence of glue.

Is it the helmet's fault for breaking? Is it the Lego company's fault for not stress testing? The fault of the child who played too roughly with her toys? The fault of the kind of society that allows a child to play? Of course not. And so, in 2009, when I arrived at Hôpital de Cery, in Switzerland, which sits atop a grassy hill on the opposite side of town from Lake Geneva, past cows roaming the bucolic hills, I knew to reserve judgment about the many, but surprisingly repeatable, ways brains can break. The research plan was simple, in a way. I was given access to the archive of art and allowed to ask: Was the artistic output of the patients related to a kind of linguistic or thought disorder, since all were just kinds of movement? Did certain kinds of mental illness lead to certain kinds of brushstrokes or tableaux?

The day I arrived was the last day of an art exhibition put on by the patients in which the hospital basement—dirt, brick, must, and soot, like a Victorian chimney with its walls set back—was converted into a gallery. One of, if not the first art piece I encountered was titled, in French, something like "This Is What It Is Like to Have Schizophrenia." It took up a whole room. It had inside of it, wrapped in brick and supported by dirt, a table, a toy train, and a mirror. On the table, the train looped slowly around the oval track.

As I walked forward, I approached from the doorway but could not see my reflection in the mirror. Strangely, only the toy train and the room were reflected. The mirror, straight out of a dark German fairy tale, seemed to have chosen to accurately reflect everything but me, as if there was a small fissure in time and space and only I had fallen in. A few seconds later, I suddenly appeared—it was my past self walking up to the table from a few seconds earlier. I moved my arm up and down, but mirror-image me, again, did not move until

Endeavoring to Grow Wings

Perhaps the human species was endeavoring to grow wings.

—**Frigyes Karinthy,** *A Journey Round My Skull*

In 2019, Dr. Jonathan Leong—a neuroscientist with a Ph.D. and M.D. from Stanford, and at the time a radiology resident at Harvard—passed away from complications of a brain tumor at age thirty-seven. Years prior, while he was in graduate school, a brain scan after a fall from a bicycle revealed a small tumor in a part of his brain called the cerebellum, which hangs off the back-bottom, near the top of the neck. Over the years, the tumor grew.

Months before the conversation below, Jonathan had approximately nineteen billion neurons surgically removed from his cerebellum. Nineteen billion is a staggering number—about one-quarter of the brain's total number of neurons. And yet after the surgery, he seemed just fine—articulate, vibrant, and, according to his own report, all still there in all the ways that counted. In similar surgeries where brain tissue is resected, or removed—common for cancers or in cases of intractable epilepsy—questions of consciousness, free will, and identity are at the center of both living with the disease and the recovery.

What would a neuroscientist like Jonathan notice or think about the removal of one-quarter of the neurons from his brain? Did the removal of all these neurons change his consciousness in any way? How could it not? From the outside, how would we ever know? From the inside, how could he?

The conversation below is between me, Jonathan, and Christof Koch, a consciousness scholar and neuroscientist. I first met Koch in India, at a conference where he debated theories of consciousness with the Dalai Lama (see chapter 14). Decades prior, Koch had also, in a paper coauthored with the Nobel laureate Francis Crick, introduced the phrase "neural correlates of consciousness," or NCC, into consciousness research. NCCs are the minimal set of events in the brain needed to give rise to a noticeable change, of any kind, in conscious or subjective awareness. If, while floating on your stomach in the open ocean, you see a silhouette coming at you and your mind's eye switches the perception between a small fish close by or a large fish far away, there is an NCC in your brain that should uniquely correspond to your conscious experience of each.

Many questions about NCCs remain. Why is there something that it feels like to be someone who has

NCCs? Where are they? Is there one for every possible experience?

Before the three of us spoke, we all read the Hungarian writer Frigyes Karinthy's memoir *A Journey Round My Skull,* published in 1936.[1] Karinthy had a tumor the size of a pomegranate in his cerebellum and, like Jonathan, had it removed a few years before ultimately succumbing to complications. In his memoir, Karinthy contemplated whether removal of tissue could counterintuitively leave one with more function than when one started. The idea is mostly literary, almost poetic, but not unheard of in biology. Mammals have many vestigial genes and traits that could be covered up in modern humans, and resection of tissue could be the removal of an inhibition, like taking the safety off a gun.

CK: How are you doing, Jonathan?

JL: Things are, I think, going pretty well. Pretty stable right now. I just turned in my last bit of scans at the end of December. There was just a very subtle change in the enhancement pattern, so we're just watching that at the moment. I completed chemo, and I'm just waiting.

CK: Your tumor is where? On the lower side of the cerebellum? Next to the fourth ventricle?

JL: Yeah. It's more in the fourth ventricle and the brainstem, and less cerebellar hemisphere. That's where it used to be.

CK: The surgeons took out thirty-six square cubic centimeters? That's a large chunk of tissue.

JL: A pretty big piece of brain came out, yeah.

PH: How many neurons?

JL: A lot.

CK: When was the operation?

JL: It was in March 2017.

CK: Has there been a change in your experience of the world?

JL: There are certain changes that are secondary to the resection. For example, I have a stable diplopia in my left eye. And I have a weird nystagmus, due to lesions in the cerebellopontine angle. But aside from these, sort of, more mechanistically explainable things, I feel like myself, actually. I don't think that consciousness has really been affected by my tumor.

CK: Apart from the nystagmus and the diplopia, the world doesn't look different? It doesn't smell different? Your introspection doesn't change somehow?

JL: No. As far as I can tell. I have the same personality. Nothing major or overt.

PH: On some level, is that because of the shallowness of the cognitive tests? The neurologists aren't inquiring about your internal, subjective, personal traits, right? They're not inquiring about identity.

JL: That has something to do with it. We haven't

really probed very carefully or very critically what is different about me. It's just my subjective sense that I'm reporting.

CK: Which is the primary data that matters. Because, after all, my consciousness is intrinsic to me. More important than the external observer is the first-person perspective. In the annals of anosognosia,[2] famously, people can be completely unaware that they're blind. They believe they can see but they're actually blind. Do you have some sort of anosognosia for certain specifics of consciousness?

JL: That's sort of a difficult question for me to answer.

CK: That would require third-person validation, correct.

PH: One can never answer that precise question. It's one of those impossible "Tell me the things you're not aware of" questions.

CK: From direct experience, no, you can't do that. I'm not aware of my blood pressure, but I know of it through a third-person perspective because people have done the experiments and I can measure it. And I think the same would apply for all these cases of anosognosia. You can infer there's something missing but only indirectly. But you, Jonathan, you don't think that it's the case for you at all.

JL: No, I don't think so. No. I feel very lucky that the "me" part of me is still intact.

PH: So, then what do you think those neurons were doing?

JL: That's a really good question.

PH: Do you guys know Plato's thought experiment, the ship of Theseus? A famous wooden ship stored in a museum, and all the planks are replaced one by one and the question is, at the end, is it the same ship? This seems related. The metaphor here would be that maybe we've removed a tiny patch on one of the sails. The ship still floats. It looks like a ship. It acts like a ship. It is a ship, but maybe it's slightly less efficient at picking up wind? All those neurons had to be doing something, right?

CK: Well, Jonathan, have you tried to learn any new sports or new motor skills since March?

JL: That's difficult now. Even old skills that I used to be good at, they're much harder for me. I think this is a little bit difficult to implicate cerebellar function, or cerebellar dysfunction, because learning new skills may be more difficult, for example, because of my dysmetria. Or it might be more difficult because I'm lacking the function that the cerebellum would otherwise serve.

CK: So what skills? Can you say something about it? What skills are more difficult now for you?

JL: Playing piano. Typing. Things like that are the things that I notice the most.

CK: So you're playing less well now, the piano, or less

fluid, or the cadence is off, than you used to?

JL: Yeah.

CK: This is a classical function of the cerebellum, right? Tempo. Synchrony. Choreography. Effortless motion. Execution of flow. There's some of your answer, Patrick.

PH: I was recently in a debate about the technical hardware and software limitations that are preventing us from being able to upload consciousness into a computer. And so these people are like, "Oh, what we should do is we should take a brain and cut it up into tiny little pieces and we'll do scans and we'll figure out the structure of the thing and then we'll just upload it. And then we'll use that virtual human to train artificial general intelligence, or general artificial intelligence, and everything will be great." One of my questions was, Why upload the whole brain if most of it doesn't matter? Why upload the cerebellum, for example, if it only coordinates motor learning? That's going to be a ton of work for this AI creature that doesn't even have any motor needs. It lives in the cloud, not in the trees of Africa like a primate.

JL: We keep pointing to experiments where we carve out some giant part of the brain and the person is fine or there's no obvious defect. And this means that this part of the brain is not necessary for whatever function that was being tested. But is that framework of reasoning going to be enough for us to be able to explain where consciousness is in the brain? I don't know.

CK: Ultimately we need a theory of consciousness. We need a theory that tells us why particular bits and pieces of highly excitable matter give rise to consciousness. My liver consists of a few hundred billion complex cells of maybe a hundred different types of liver cells. But they don't seem to constitute consciousness. Why is that? Where's the difference between the brain and the liver in terms of physical systems?

PH: Jonathan, do you know the IIT theory?

JL: I've read a little bit.

CK: So let's step back. Integrated information theory infers the physical substrate of consciousness. It computes this number called phi of all possible networks in the brain. In fact, inside the entire body. The physical substrate of conscious experience is the network with the maximum of the integrated information. Here, the argument would be the cerebellum simply isn't part of that special network because it has a very low phi. The cerebellum is organized in these parallel circuits, each one of which is primarily wired up in a feed-forward manner. Very, very different from the wiring in the neocortex. Strong evidence suggests a posterior hot zone somewhere in the parietal/temporal/occipital cortex at the physical substrate of human consciousness. In principle,

my phi would not change at all if you remove my
cerebellum.

PH: What about when I imagine things? When I
daydream about the future? When I do that, I
imagine myself and sometimes I imagine moving
through the world. My visual imagination is a
motor process.

CK: Maybe imagination does activate some aspect of
cerebellum, but the question is, Is that conscious
or is that unconscious? Even if it's unconscious, it
may contribute. There are lots of things in my body
that contribute to my consciousness that I don't
have direct access to. I know, for instance, that
in order to see depth, my brain needs to compare
the left eye and the right eye. Consciously, I have
no access to that information. I simply do not
see separate left and right images. I only see this
integrated, three-dimensional view. So, yes, I
would say the cerebellum may be partially involved
in motor imagery and in mental time travel, but
unless it directly contributed to consciousness,
it just serves as a background condition. Your
conscious sensation, your conscious experience
wouldn't change much without your cerebellum.

PH: So consciousness is different from the richness
of said consciousness?

CK: Tell me, Jonathan, do you experience mental
time travel, mental imagery, motor imagery?

JL: What do you mean?

CK: Can you travel in time back and forth?[3] Can
   you imagine? Tell me something about the day
   you're going to get married. Tell me something
   about what you will do when you have a daughter.
   Those are questions in principle you should be
   able to answer, "Well, if I have a daughter I would
   do this and that with her. I would take her here
   and here and the other." Some people can't do
   that, particularly if they have hippocampal lesions.
   Turns out these people not only have retrograde
   amnesia but also have difficulties traveling forward
   in time, imaging a future for themselves.

JL: No, I feel like that is intact in me. I can
   anticipate. I can make plans for the future and far
   into the future. In addition to going back into the
   past.

PH: When you remember playing the piano, or
   doing a complicated motor task, which you were
   once fully capable of, you conjure that up in
   the present, do some of the motor coordination
   deficits that you encounter today, do those go back
   with you into memory?

JL: The question being, if I remember that time
   or those events, is there some change in how I
   perform now, is that your question?

PH: Yeah, does the deficit go back in time with you?
   I guess I'm trying to differentiate between two
   models: One, when you instantiate a memory,
   it fully simulates a brain with eighty-six billion

neurons. It remembers full and complete feedback and the exact motor movement. And the other would be, this seems more efficient, to, when you instantiate a memory you take who you are now and simulate being back in that environment. But if that were true, you might imagine that when you simulate it, you have the motor difficulties that you currently have, right?

JL: I can remember being good at things and my deficits don't follow me backwards or forwards.

CK: That's fascinating.

JL: Or maybe I'm just remembering the outcome of what it's like to have the whole brain. I don't know.

PH: Do you think flies have consciousness?

JL: I hope so, yeah.

PH: Did you find yourself believing they were more conscious as you studied them?

JL: If anything, I realized they were less and less conscious.

PH: What do you mean by that?

JL: Well, I think there's certain functional definitions that you could use. For example, they don't sleep the same that we sleep. They don't solve problems in a social way. So, these things aren't about consciousness, perspective say, but they're kind of shadows of it.

CK: Wait, but they do sleep, right?

JL: Yeah, they do sleep but it's different than what we do, right? Whatever consciousness they have, it's not quite the same as what we think we have.

PH: Do you see a relationship between consciousness and free will? In learning more and more about the fly, did you find them to be more and more deterministic and therefore less and less conscious?

JL: Yes. That pretty much sums it up.

CK: So, wait. No, wait, wait. How can that be? Fly scientists have done these experiments in which one fly at a time is placed inside a labyrinth with branch points where it has to either turn left or right. For any particular fly, it's impossible to predict whether on this trial it will turn left, or right, or will sit still, unable to make up its mind, or perhaps even turn back. Statistically you can predict these choices but not for any one animal at any one time. To an observer, it sure looks like the animal is freely deciding what to do.

PH: There's this adage that the difference between physics and biology is the difference between dropping a bowling ball or a pigeon off a tower. "Okay, you drop a pigeon. Does it fly left or does it fly right?" When we're searching for a unified theory of consciousness, it should be able to explain everything in a way that's predictive, right? Including the pigeon. We can slingshot a satellite into the cosmos and have it hit an asteroid with millimeter precision because our theory of gravity is so good at prediction.

CK: The world is deterministic at a very large scale. Not at the microscale. So, whether that works for consciousness or not, we simply don't know at this point.

PH: Jonathan, is there any sense in which, by removing all these neurons, that you feel in any way whatsoever that you're more deterministic?

JL: I don't know. That's a good question. I guess the corollary question is, Would I even know if I were more or less deterministic? I don't know. I don't think that I've seen any changes in my behavior to suggest that I'm more or less deterministic.

The North African Rhino of Charismatic Megaquale

Many of those interested in animal conservation have a bias toward "charismatic megafauna"—that is, those animals that are large, blunt, and beautiful. The blue whale, the North African rhinoceros, the giraffe, and so on; algae, viruses, beetles, and ants get short shrift. However, mass and charm alone are not what make nature go round. If the absolute numbers, diversity, or uniqueness of coding genes were the highest conservation goals, perhaps each of the thirty thousand beetle species or the trillion-plus species of microbes on Earth should be more than worth protecting as all rhino, whale, mountain gorilla, elephant, lion, or polar bear species combined.

So, too, do many theories about how a brain works focus on what could be called "charismatic megaquale"[1]—that is, those experiences that feel big and vibrant and can charm us into thinking that there is something it is like to be something: what it is like to be a bat, what it is like to be a North African rhino, what it is like to try to remember how many windows were on your childhood home, what it is like to remember looking through your own eyes right now, and what it is like to experience large, generous emotions like fear, attraction, happiness, and sadness, which seem to take over the subjective world in one large brushstroke, coloring all its objects. But so much more is happening to a brain in the process of acting on itself. Populating consciousness are more species of noncharismatic minor quale, which are often ignored and undetected, than any other kind. What is the point of theories of consciousness if they cannot just as easily explain the minor quale of human emotion as the megaquale?

The first person whose prose was ever called a "stream of consciousness" objected to the phrase so strongly that she wrote a letter to the editor at *The Egoist* magazine saying that consciousness "sits stiller than a tree."[2] It's really not just one thing because the human brain, like Earth, formed on geologic time. A search for consciousness through the skull, like one for minerals, elements, or ore through Earth's crust, is made more efficient by knowing the natural conditions required for its creation. A search for surface gold, for example, is made easier by knowing its surface correlates, called alluvial deposits, which result when the flow of water over millions of years concentrates and separates everything it has washed over. So, too, the search for surface diamonds is made easier by looking for volcanic pipes that bring diamond-bearing rock to the surface; or in the search for helium, where one hopes for geothermal pools above a rock of radioactive elements, meaning they are likely to have been cooled and cooked for a very long time.

Are there likewise a set of physical conditions or surface correlates that, if found, could point to where consciousness is in a brain? If you split Earth in half vertically, along the prime meridian line, and separated the halves so that gravity would not immediately force them together again, Earth's iron-nickel core, relieved of the pressure that keeps it solid or molten, would explode as a gas. The atmosphere and oceans would drain and the magnetic field would fracture, allowing solar wind to strip away the ozone.[3] Because each of the two halves would not be a perfect sphere—meaning their centers of mass would no longer be at their exact center—the two planet halves would either stop or increase their rotations into a wild spin, causing earthquakes and tsunamis and the end of all life as we know it.

However, strangely, if you fully cut the cables connecting the two hemispheres of a human brain, which is the most complex physical object in the known universe, it will leave two minds and not two halves of one mind behind.[4] One interesting implication of this is that such an experiment should, in principle, reverse. There is no reason to believe two intact hemispheres could not, with enough cabling, be connected to a third, fourth, seventeenth, or one-millionth additional hemisphere. This, were it to happen to you, would not make you feel substantially different from the way you feel now; perhaps, at first, a little less predictable. If your two current hemispheres were two of the million cabled together, you wouldn't exist anymore, per se. Though no modern technical or medical definition of death would label you as "dead," you would have died the moment a cable was connected that changed where the hot spot of self-connectivity was greatest. Subjectively, you would neither notice nor remember your own death. As aerosol particles in a room spread evenly between its walls, consciousness also expands to fill any room it is given.

Consciousness, like SPF, IQ, and blood pressure, can be reduced from a complex set of measurements to a simple number.[5] As an expert prospector should be able to stand at the highest spot atop any horizon and point to where gold might be, consciousness prospectors should be able to take any network of connected things and point out where the hot spot of consciousness should be based on its terrain. The analog to the brain's kimberlite is clear; it is the spot in a network that most acts on itself.

In 2016, I was invited to fly to India and watch a debate on modern theories of consciousness hosted by the Dalai Lama, the spiritual leader of the Tibetan people. If the country border of India outlined a baseball diamond, the monastery is where a base runner, already having rounded third base, would start their slide into home plate. The monks who call this temporary home theirs, being outcasts from Tibet, were given the India equivalent of nineteenth-century Oklahoma: barely arable land, far from the sparkling coasts. The balcony to my room opened onto a wall topped by rusted barbed wire, with years of plastic and trash stuck to its thorns, alit there by the wind. On the other side of the wall were oxen and subsistence farmers, in grainy silhouette. Seen from above, the walls of the monastery outline a tract of land that resembles, because it is closed off, and because it is ovular, a cell.

The wall and wire fence, like any cell wall, are a kind of membrane with properties like batteries. Electricity is just the separation of charge, charge that falls naturally across any gradient, charge that you can see built up at any country borders or any railroad track that splits a town in two or at the border of any neuron desperately using its sugars to keep the outside selectively out. I can still see, at scale and in my mind's eye, a bird's-eye view of events that week; can see myself and the neuroscientist Christof Koch standing in the town around the monastery, pointing into the horizon through the crop-fire smoke, as tourists might, at objects in hazy relief as we questioned how much of the visual scene the brain makes up versus detects all while mostly ignoring the oxen and poverty and the plastic bags caught like confetti in the barbed wire.

Over dinner on the first night, Team West, composed of invited professors from various prestigious U.S. institutions, debated whether it was okay to kill mosquitoes in front of the monks (no) and whether there was Wi-Fi (yes). The rest of the debates, a friendly battle of east versus west for cosmology, ethics, the self, and consciousness, were to take place the next day in a large temple. While staring at the temple roof, I imagined an old saying that I liked, that the difference between biology and physics is the difference between dropping a pigeon and a bowling ball from high atop a tower. The bowling ball drops straight down, subject to nature's forces, of course. Newton told us that. The pigeon, though, if alive and able, flies away to survive. Darwin told us that. If we imagine the pigeon and ball the exact same weight, all of biology can be reduced to a simple question between the two: What, exactly, is different about the configuration of atoms in the pigeon that gives it the freedom to break the pull of gravity, flap its wings, and rise against the air in a way the bowling ball cannot?

The answer is in the properties of its network of cells called neurons, which are simply cells with walls and parts and metabolism and health and DNA and organelles, just like all the others. The bowling ball has some sort of physical, regular structure to it, of course, or else it would not be definable as what it is. A theory of consciousness or a consciousness "meter," like a dowsing rod, should be just as capable of explaining why the bowling ball *is not* conscious as why the pigeon *is*. The measure of how conscious the pigeon is can be determined by the maximum value of its so-called phi, or the symbol $\Phi_{(max)}$, which stands for the intrinsic cause-effect power of the local maximum of cause-effect power—which is, in other words, the ability of any group of communicating things to act on itself. The bowling ball, because it has some structure to it but very little ability to act on itself, has a hypothetical $\Phi_{(max)}$ of, say—we're just making this up—0.001. A plant, which has molecular feedback loops that act on themselves, has a $\Phi_{(max)}$ of, say, around 60; a pigeon, 120; a dog, 150; *H. sapiens*, 250.

Others have put forth their own treasure maps. Descartes, in the mid-seventeenth century, noted the asymmetry of the brain's lone pineal gland (inflated, he thought, by "a certain very fine air or wind" of the soul), which, by naked eye alone, appeared to not have a double. Descartes was wrong, though. It is the intrinsic structure of his brain's connections, not the quantity, type, ability, or asymmetry in any one of its neurons or regions, that gives rise to consciousness in certain parts of a brain more than others.

This explains in remarkably clean terms why splitting a brain in half results in two people rather than two halves of a person—their $\Phi_{(max)}$ resolves to two places suddenly, instead of one. It also explains how adding more hemispheres—many smaller parts, each with their own $\Phi_{(max)}$, at some point resolves to have a $\Phi_{(max)}$ across them all—would, hypothetically, result in a unified consciousness. Removing a piece of brain useless for consciousness, like tens of billions of neurons from the cerebellum, say, would be like digging for treasure in the wrong spot. It would not change $\Phi_{(max)}$ at all.

The terms of the debate were set. The fact that insects, birds, and bats can fly should not motivate a single explanation describing how they do—insects, birds, and bats all evolved for flight separately, each cobbled together with pieces from different evolutionary sets. It would be misleading to find order in their respective abilities, and yet biologists have a history of being tricked by visual similarity. But should the fact that insects, primates, and the like exhibit displays of consciousness mean that they all do so in the same way? (Is this why it's okay to step on a Wi-Fi router but not a fly?)

The morning of the debate, with low-Φ mosquitoes greeting the dawn, my blood half tea/half Malarone, the flow of my thoughts started to go unexpectedly, in a way that it hadn't ever before. Still, I walked with Koch to the temple, took off my shoes, and floated barefoot to my sideline seat a few rows back from where the guests would speak. I couldn't control where my mind went. *Will there, one day, be a physics of thought?* Physicists can softly land a car-sized robot on Mars and detect the spectral origins of the Big Bang, but we in the neuroscience world cannot even explain what panic is. *Why can't I keep still? Or stop touching my face, like there was a woolly blanket under my skin?* Imagined as a landscape, any given thought's speed and content is determined by the underlying topography of memory or can be roughly predictable in the same way that a volcano's lava flow can be roughly predicted by the flow of its snowmelt.

So, thoughts go somewhere, one after the other. But why there and not elsewhere? The sound of three thousand monks sitting is like the sound of a monsoon hitting a shingled roof. *Why these thoughts and not others? Didn't I just wonder that?* I was going to get up and leave, despite the eyes of thousands of sitting monks on me as I walked out, each of them obviously at inner peace with themselves and their bodies and minds while here I was, antispiritual, anti-Buddhist, my blood turning into a boiling tide pool and a phantom limb speared through the left half of my ribs. The statistics of spontaneous thought may be broadly similar to statistical models of foraging behavior in natural environments, which can mathematically describe goal-directed movements in a variety of organisms including single-celled parasites, immune cells, herons, bees, and humans attending county fairs. *If the goal of the brain is to predict, and prediction means silencing, do bodies disappear when their goal is done? When they've predicted everything? That can't be right.* But a perfectly predicted world vanishes from the mind's eye, right, compressed like a television signal? So why was each part of me so suddenly there? *Maybe if I just cut the whole thing off at the neck.*[6]

Foraging creatures spend a large percentage of their time searching in their immediate surroundings ("perchings") with an occasional burst to a new area ("flights"), and the mathematical ratio that describes this movement, called a Levy flight, or Levy walk, is broadly consistent across organism and scale. *The monks all seem* a feature of the physical terrain, the place where two rivers meet. But for those, like me, who did not grow up near a river, I didn't know that the way out of a whirl-pool is not to swim against it but rather to give in and let it take you around a full loop to the top, upriver, but where escape is easiest.

Outside, the balcony stairs were now covered in three thousand pairs of monk shoes, and I went about searching for mine like a lost golf ball in the rough. I was reminded during this fruitless search, with my eyes darting around, sampling the environs for reward, that India was not my first experience with Tibetan monks. When I was in graduate school, for complicated reasons involving my having tutored the son of a professor, I had been living in a spacious but mostly abandoned house in the heart of Stanford's campus, all by myself. It had five vacant bedrooms, and so when the Dalai Lama, the same one, was visiting campus, and when his attend-ing monks had refused the university's offer of a hotel room, it slowly got around that I had the extra rooms. Of course, I said yes and offered the rooms for a week all while news had spread that a dead blue whale, the most mega of megafauna, had floated ashore on the Califor-nia coast.

Through a translator, I was told that the monks had never seen an aquatic creature larger than a small fish. Could we rent a van and go see it, they asked? And so we did. By the time we got there, the whale had whit-ened, and was bloated with decay. On the way home,

15

An Itsy-Bitsy Teeny-Weeny Quantum-Dot-Like Non-Machiney

But no one knows for certain
And so it's all the same to me
I think I'll just let the mystery be

—Iris DeMent

It is often said as a criticism that theories of how the brain works latch on to whatever complicated technology or idea exists at the time. Hence, we got steam-based theories of the brain in the eighteenth century, loom-based theories in the twentieth,[1] computational theories in the twentieth, and quantum-computing-based theories in the twenty-first.

Considered in the inverse, these criticisms represent a remarkable fact about nature. It implies that no matter how complicated a discovery we make, we find out shortly afterward that nature has already both known and taken advantage of the phenomenon. Long before Newton discovered the laws of gravity, our muscles stretched in calculated attempts to defy it and our eyes tracked objects as arcs in the sky, which accelerated somewhere around ten meters per second squared. Long before we understood Earth's magnetic field lines, birds and butterflies were using them to migrate. Long before electricity was discovered, neurons were electrically signaling to each other and themselves. Long before reinforcement learning revolutionized artificial intelligence, cells were keeping physical records of their experiences. Long before Einstein explained relativity, our brains were coordinating observer-specific time frames. Long before Planck or Bohr explained the mathematics of waves and particles, retinas were noticing single photons and summating them as quantal packets of information. One can only imagine what else brains might harness from the universe as future scientific fields that we have not discovered yet reveal their clues, too.

And so, as a kind of meta-argument, or argument about arguments, why should it not be the likely expectation that when future scientists understand more about the smallest, most mysterious physics of the universe, we will also find that the brain has learned to take advantage of the most remarkable parts of these, too? Can one truly be of the belief that we have discovered it all, in the early twenty-first century? That we will never have more complicated machines than computers?

The study of the brain is in its Babylonian era of discovery, at least one thousand years behind the study of the rest of the universe. Like the star charts from the astronomer Ptolemy, almost two thousand years ago, which could nicely predict where planets would be next but not why, we know today where cellular activity in the brain will be, but not why. One reason for this is that the history of twentieth-century study of the brain is a history of "slice chauvinism." For decades, neuroscientists often looked at slices of the brain, which distort a three-dimensional object into a mostly two-dimensional one, because they had to. That is, the tools used for most of a century of study of the brain, due to technical constraints, were forced to choose a scale: glass pipettes, which stick to the outside of a cell's membrane, learned only what the membrane knew; microscope lenses, which glorify the cell body, have preferred ranges of focus; genetic stains, which can light up cells like a string of holiday lights, choose some cells and not others, but only most of the time, and never fully. What is missing is the ability to peer into the atomic scale in vivo, into the parts of cells that remain mostly unseen.

When researchers asked Anna, a teenage girl, why she laughed when a surgeon stimulated the part of her brain that caused her to laugh, did she alone will the explanation into existence for why she laughed or did the universe simply move around its atoms in a foregone way because it had to? One possibility is that the universe, since its birth, has unfolded by physical law down a path determined all the way from the relative concentration of lithium in the post–Big Bang years to the precise angle and stimulation current selected by Anna's surgeon when probing the part of her brain that then seemed to cause her to laugh. But this seems to undermine the intuitive surprise of her answers for why she laughed: "you guys are just so funny . . . standing around" and "the horse is funny."

In the way that a bright light can affront the retina and cause someone to blink or look away or in the way that a sound louder than one hundred and forty decibels can startle a person or damage their ear, Anna was both forced to laugh by the electrode in her brain and free to explain why she did.

Put another way: the two possible categories of stuff in the universe are (1) things that are determined by the physical laws of the universe and (2) those that are not. Any computational system that proceeds only in binary or ones and zeros is in the first category. That is, on some level, its functions and outputs can be predicted by its inputs or prior stats. Thus far, we have no evidence whatsoever that neurons and all the cells and molecular machines in the brain, no matter how fancy the equations are, are not fundamentally computable. Which means, if that holds, that they are thus also determined. But are we fully determined? Was Anna, in her responses? Was the surgeon, in his choice to operate or in where or which regions of Anna's brain to explore? We know from introspection and axiom that consciousness can have thoughts about itself that, by analog to Gödel's incompleteness theorem, means that the brain's computations cannot be completely described using the findings of the system itself.

Thus, the search is for some place, some hidden nook in all of life, which has noncomputable, nondetermined features.[2] Such a place cannot exist on the scale of most of life—persons, organs, cells, or enzymes—because they are too large (too noisy) and too messy (too hot) to draw anything from quantum indeterminacy, which requires an observer and a clean, noiseless place to play. Just as residents of a small town with only one well have a single source from which to draw water, the only possible place to find the "observer," whence free will and consciousness draw their power, is deep, deep down inside something called a microtubule, a paper-towel-rod-shaped thing that acts as a shield for the absolutely pure water inside it. Only from here can the observer, which is some invisible or molecular happening, pull a bucket of nondeterminism from the universe and aggregate it upward into what we then call free will.

We got here because evolution by natural selection does not operate exclusively on the large and living but over time on any process whose participants can either exist or expire. Evolution by natural selection therefore influences every single action of each molecule during all the billions of years of interactions at the level of the organism or individual. The process is then multiplied by trillions and trillions of molecular and atomic collisions every second in every cell in every living thing on Earth. If there was a mystery building block of free will hidden in the brain that, if tapped into, allowed the organism to break the emergency glass of determinism and be free, finally, and that along the way happened to give rise to consciousness, nature surely would have found it by now. Just because we have not does not mean that nature has not.

It is tempting to believe that neurons are where will happens and that their electricity is at their and thus our command, but largely this is because they show up so nicely on the slides and we invented light bulbs and electrodes and oscilloscopes at just the right time to notice, coincidentally, that the brain can be cutely electrical sometimes. But at its basic core, the brain maintains a code of on-or-off, all-or-none, which means it can be computed, which means that the solution to the mystery must be found elsewhere. There is no mystery at the scale at which cells send their messages—understanding, and therefore consciousness, must be somewhere outside the parts of the brain that merely compute.

Likely, this is in the space in cells, between the support beams of every sufficiently complicated cell. If the hope was to trace the causal forces backward in time from Anna's explanations and laugh, the only possible place to look for the font of her choices is in her microtubules, which can play host to events at the level of very tiny things. Thousands of these microtubules are inside every animal-like cell on Earth, including each of Anna's. They are the support beams that give cells their shape, their capacity for movement, and their strength. Within each microtubule, which is around twenty-five nanometers across, is the only place in the human brain that is noncomputable.[3] They are about three hundred times as large as the space between bonded atoms or one-three-thousandth the width of a human hair.

Inside each microtubule, which are hollow and aqueous, like waterslides, our ability to predict physical interactions ends. Interactions at this scale are either probabilistic or random, though we do not yet know which. Here consciousness arises, as small units of proto-consciousness, in the last known place for them to be. They are there because they have to be. They cannot be at any larger level because groups of cells or brain regions are much too big and warm and messy and can be at least hypothetically, one day, simulated by equations. The only place for consciousness to draw its powers is from inside microtubules, because within each is the only place in the brain clean and isolated from the outside enough to tap into the surprise of indeterminacy.

Here, and only here, we find the unknown parts of physics and mathematics, which still cannot explain what happens or why it is the case that observing something changes it. Nowhere else in any brain is it clean, cold, shielded, and observerless; the only feasible place, thus, is where our knowledge of both physics and biology ends and where our knowledge of Anna must begin.[4]

A Make-Believe Parasite with No Legs and Places to Go

Picture the human brain as a vast manufactory, in which thousands of looms, of complex and differing patterns, are habitually at work.

—**Frederic Myers**

Though many of the descendants of Ernest Hemingway's polydactyl cats also have six toes, there is no reason they, or any cat, could not have six thousand toes.[1] The brain of such a cat would find a way to control all of them and its consciousness would do its best to have movement and coordination plans learned and at the ready. There is no known physical limit to how much space a single consciousness can "inhabit."

Clearly, consciousness comes in different sizes. A mouse, a human, and an elephant are all controlled by a brain that extends the edges of its control to the edges of the respective body that houses it. If a tiny mouse brain grew in an enormous elephant's body, it would learn to control all the parts of the elephant all the same. Whatever the largest animal in the history of life on Earth is, it had a brain that brought subjectivity, pain, and a concept of self to its borders, too. Those who disagree with the claim that consciousness can be infinitely large have the burden of proof to show that there is a limit to the space a single consciousness can inhabit. What is that limit? If there is one, why? Is it a physical limit? A thermodynamic limit? Could we imagine a galaxy-wide consciousness, as long as a certain speed was maintained?

As far as we know, and as long as someone figured out the plumbing, a conscious creature of any size is consistent with known properties of both physics and neuroscience. Its consciousness would envelop whatever body and all the toes it had control over because a brain's only output is focused through a few neurons reaching out into the world. This must be the ultimate purpose of consciousness: to control a body. If half of the polydactyl cat's six thousand toes were to disappear somehow, there is also no reason to believe the cat would not feel three thousand phantom toes in the approximate shape of their absence. Consciousness, like a gas, fills to any volume it is given.

We know this works in principle because, in the mid-twentieth century, it was shown that a human brain could be split in half by cutting the two hundred million wires connecting its two hemispheres. Those with complicated seizures, and as an alternative to removing the brain tissue at the seizure's source, could take the more drastic measure and have the cabling cut so that the electrical storm stayed localized to only one half of the brain. Remarkably, the two halves each then showed signs of independent consciousness, with each hemisphere having distinct abilities (in some cases, for example, only the left hemisphere could speak), preferences, and memories.

Instead of halving consciousness, one could imagine the reverse: an operation where two otherwise normal brains were connected with adequate bandwidth, in a process called "bridging."[2] Would the combined brains form a single, subjective conscious mind or remain as two?

Why is there any reason to presume that the natural bridge that we all are born with, which connects our two hemispheres, could not be expanded upon? Bridging would directly connect billions of neurons in one brain with those in a second, mimicking the brain's natural bridge between its two halves. Two people, once bridged, would be able to share all of their sensations, daydreams, memories, and thoughts. They will respond to questions about their experience as if they are a single, unified self. Just as we do, as a single voice, despite our two hemispheres.

Consciousness can be shaped to anybody it has direct sway over. However, these thought experiments undersell the fact that we do not have a good definition or boundary line for what even a single mind is. The vagaries of a day, whether one is hot or cold or has had their coffee or remembered a joyful thing or is in diabetic shock or is infected with a brain or leg parasite or the flu or is running from a lion or in a conference call or locked out of their postal box or slept poorly the night before, changes the speed and content of conscious and spontaneous thought.

As well, some people claim to have no internal visual imagery; some see images in their head as a flickering slide reel; others as if the memory is happening to them from the vantage exactly as it once did; others watch their memories unfold from a few dozen meters away, through a single aperture, as if filming them; others as if they are watching a television one hundred feet away. Some have photograph-like memories and can redraw cityscapes from scratch. Some people claim to have no inner, vocal monologue. Some can remember the details of every day of their lives and exquisite timing details about the details. Some, while reading this text, see each letter as colored, and the colors align mostly with the Fisher-Price magnetic alphabet set they had as children.[3] Some infer based on their experience with outside or inside light that objects, like dresses, are differently colored. This immense possible variation in conscious experience is remarkable. It means that all of the experimental and the control conditions for the study of consciousness already exist across people.

Why is it the case that there is not one kind of "what it is like" to be human?

If schizophrenia, aphantasia, and variations in inner experience can be passed down from parent to child, two conclusions can be made. First, that the content and quality of consciousness is heritable. Second, that consciousness is equally as subject to the pressures of Darwinian evolution as, say, caribou antlers. Consciousness is just as varied, or perhaps even more so, than the number of species of life on the planet, and though there may be broad, geologic swaths of similarity between kinds of thoughts, it remains the case that, just as every forest is different in its details, so, too, is every brain different in its.

Natural selection, though, has to act on something. It is therefore useful to consider the natural goals of consciousness in terms of the natural goals of other creatures. Like all single-celled parasites, for example, consciousness has no legs but lots of places to be. So, too, does consciousness require a host with any means to move to have any chance, achieved by invading a body and spreading outward in time to each of its nerve endings. Because any brain's only output is the motor neuron, consciousness must do only one thing well: reduce the cost of movement by making the world appear simple and actable. To do so, the brain makes movements more efficient through prediction of what movements will do based on everything that has been tried before; it makes a vague model of the outlines of the body to aid in unplanned, surprising demands or unexpected reaches. What falls out of these compression steps is the feeling of what it is like to be something. This something is dependent each and every time, within each and every brain, on its setup.

An obligate parasite is one that depends on a host for survival and reproduction.[4] The classic story is that of a virus that requires the machinery of a host cell many times its size to make more copies of the virus. But humans, too, require oxygen from the air and a handful of amino acids in our diet, which we cannot make on our own. Without them, we die. Are we then not, in some sense, also parasites of our surrounding environments, required to strip oxygen from our planetary host? We parasitize our local environments, stripping them for parts to mix with carbon.

Likewise, our bodies host a parasite tucked in between every atom's mass in an impenetrable, nuclear fortress.[5] It is a rather bizarre feature of nature that when two humans combine to make an eventual new consciousness, that process resets as a single cell. A naive engineer, perhaps, would design the reproductive step of creating more consciousness to be like budding off an existing brain, as one might with a mushroom, fungus, or plant graft. Instead, we can pass down two combined versions of consciousness through the filter of a single wired cell. Which means that the single cell contains within it all the rules to one day become conscious. Like a drop of rain, which requires a nucleation event to form into a liquid, consciousness needs a brain to form around.[6]

In a future world that allows bridging, when brain-machine and brain-brain interfaces are common and when neurons can be connected like yarn stitched together, what new patterns of conscious activity can be made? What if you crossed species? Could you learn the tiniest bit about what it is like to be a bat? If all language learning happened because of a moment in evolution when a few restitchings of hand gestures mapped onto the larynx, what other endless possibilities remain? Is language just the start? What would happen if you connected parts of the visual cortex to your own, or someone else's, brainstem? Could you see your own unconscious or gain access to someone else's memories? There are eighty-six billion neurons in the human brain. If each of them can be resewn and if the body can be as large or as small or as polydactyl as desired, are there infinite numbers of "what it is like to be" something? If so, do we expect any one theory to explain them all?

A Sex-Starved Cricket
Sculpting in Time

When all was finished, it cannot be denied that this
work has carried off the palm from all other statues,
modern or ancient, Greek or Latin; no other artwork is
equal to it in any respect, with such just proportion,
beauty and excellence did Michelangelo finish it.

—Giorgio Vasari

Within every raw block of marble of considerable size and quality is a statue of the *David* just waiting for someone to come along and chip away the unnecessary bits. So, too, does a brain create a story of self as a kind of carving out from the raw sensory stream of electromagnetic radiation and physical pressure bombarding the body at all times. From this totality of incoming input we expertly carve out the relevance of light and sound and call it "seeing" and "hearing"; from touch and its coordination with seeing and hearing, along with our internal compass and the carpenter's level of the inner ear, we also carve a shape in a human form that feels to each of us like our own body's outline.

In the 1990s, a teenager, Anna, sat awake in an operating chair as her skull was peeled back and the brain surgeons got to work. Before the full surgery was done, which involved removing the piece of Anna's brain where the seizures began, and in order both to get a layout of her brain and learn the tiniest fraction more about the workings of all of us, the surgeons probed her brain in eighty-five different spots with an electrode to see what would happen. Because the brain itself does not have pain receptors, because there is nowhere left to send the message, Anna lay painlessly with a solid steel stereotaxic rig keeping her head still and a curtain around her forehead as the surgeon, out of sight, slowly peeled back her brain's many coverings.

Head fixated. Skull open. Eyes open. Hearing all the sounds of the operating room. What does Anna see? Her brain could have evolved to naturally capture other kinds of information in the environment, like ultraviolet waves, gamma waves, magnetic fields, infrared light, seismic activity, or even Wi-Fi, but it did not. Somewhere along the line from cell to primate, a selection of the relevant kinds of information from the raw streams of all inputs was made for us by our genes and habitat and social needs out of the near infinity of other kinds of possible outlines.

With her head perfectly still, Anna could move very little above her neck except her eyes, which were moving three times a second in rapid, quick bursts around the room, and her jaw and facial muscles, which she mostly used for speaking, laughing, and smiling. If the muscles that controlled her eye movements were anesthetized such that they, too, could not move, everything in front of her would eventually just simply disappear in a puff of subjective smoke. Her brain needs the movement of the eyes and head to make sense of the world. Seeing is an active process, as different from the passivity of a camera lens as carving a statue is to dropping a block of marble on a scalpel.

In other words, because she is awake, Anna is moving with her eyes to keep the stability of her mind's eye. If she closed one eye and focused on something in the room and looked just a little to the left and then right, the room would appear to stay stable. This works because the part of her brain that sent the message to pull the muscles in her eyes also sent a message to another part of the brain, which receives input from the eyes, that they were about to move. A very similar message was sent to the part of her brain that stitches together the visual world, a kind of carbon copy of the first message.[1] Thus, every time her eyes moved, even just a little, her motor cortex was involved in sending two commands through the brain's cables:

1. Eye muscles, do whatever you do to make the eyes move left.
2. Visual system and all parts therein, we just told the eyes to move left a little; please do not be surprised by the jump.

We know the brain is doing something active with its eye movements because there is another, cruder way of moving the eyeballs, which every reader is welcome to try: by force.

If Anna had closed one eye and gently took her finger and pushed, in an upward motion, on the side of the open eye, and (gently) outward from the edge of the eyeball and then (gently) up, everything she saw would rotate a few degrees, as if the room were spinning on a prize wheel. In this scenario, the eyeball has moved but without the message or its carbon copy. No messages are sent. Her visual system, which is used to always receiving some kind of warning, panics without the warning. Her visual field spins because the storytelling triptych of the eye, the brain, and its messages becomes unspooled. This happens because it is useful for the brain and body, when either moves, to know that it will soon move because it is otherwise impossible to know whether the outside world moved or whether the eye did.

Thus, every time the eye moves, the brain is actively doing something to make sure that the visual world does not subjectively jump about. That is the default way of things. The eye is not passively perceiving light as it comes in but is actively crafting a narrative at least three times a second. Every time it moves, it warns. All so that things look like they remain where they are. (One can try this also with a camera, by taking photos of direct, line-of-sight points of focus as an eye scans a room. Put next to each other, one realizes quickly that the eyes get only snapshots and that the composite must be stitched to make a simple, coherent whole.)

If this sounds fancy, it is not. A simpler version also happens with each high-pitched chirp of a sex-starved male cricket.[2] The cricket is forced to rub its wings together to make come-hither sounds so loud that their "ears," which are just little holes on their legs, should be deafened by the noise but do not, in fact, go deaf. The ball of neurons that tells the cricket wings to move also tells the ear that it has just made a sound and to silence it out.

If one was trying to solve the chirping problem, in nature, over the course of hundreds of millions of years of insect evolution, one could imagine a few ways of accomplishing this. First, the cricket might try silencing all of its hearing while chirping, so as not to damage its "ears," by sending two messages:

1. Leg muscles, rub together.
2. Spiracle holes that I hear through, I'm about to chirp, so please mute *all* sounds.

Which, of course, could be a disaster for survival, as predators would quickly learn to pounce, during chirps, on all the deaf crickets. In reality, the cricket has a subtler version specific to the expected sounds produced:

1. Leg muscles, rub together.
2. Spiracle holes, I'm about to chirp at exactly 100 db at 30k Hz. Stop listening to just that frequency range but keep all the other channels open.

The cricket's ears can silence the exact sequence of what they expect to hear based on a lifetime of experience-making chirps, and these copies act as both prediction and warning to prepare the sensory parts of the cricket for what to expect. Fast-forward up the web of life—to mammals, then primates—and we are not so different. What is language but a series of elaborate chirps?

When Anna answered the surgeon's request with "the horse is funny" and "you guys are just so funny . . . standing around," she heard her own voice. As she spoke, a part of her brain canceled out the acoustic sounds it expected to hear as her voice bounced back into her ears. Thus, we experience things that are ours as ours, including the voices in our heads, which are just motor commands, like chirps, not because they intrinsically are "our" thoughts or voices but rather only and simply because they match.

From the total of many of these matches per second, we get the feeling of subjective, first-person action and ourselves as its fluid, graceful center. These signals are common in mammals and can vary in ability. Bats echolocate into themselves using a high-end version of the trick by making their own sonar chirps and learning, over time, what the echoes mean; a small, electric fish off the coast of Africa that pulses electricity into the surrounding water uses similar self-canceling tricks with electrical currents it passes through the ocean water and across its own muscles.

What this means is that eyes are not just watching but also predicting the future.[3] Much can be learned by watching where the eye looks. If she were an expert baseball player at bat, she would look at the pitcher's elbow or release point and then, around one hundred and fifty milliseconds later, at the ball or at a predicted point somewhere the ball may cross; amateurs look askance, at the pitcher's head or the ball itself. If she were making tea, she would move her eyes to the next item she wanted to handle—pot, tea bag, teacup—half a second before finishing with the previous one.[4] Where her eyes go, her hands soon follow.

Our eyes are already experts at the small, mundane tasks we most engage in because the mammalian brain is wired for motor efficiency and every adult is a highly skilled aesthetic expert, just as good at carving out attention while making tea as any sporting great is at their sport. What appear to us the simplest of everyday tasks are anything but. Behind it all, the brain is heaving. Anna's brain is not recording the present or rehashing the past but, even while still, is constantly making the future. For the sake of efficiency, the brain prioritizes newness for these predictions.[5]

The *David*'s original white marble block, cut from an Italian mountain, was eighteen feet tall and weighed more than twenty-five thousand pounds. At the time, it was the largest quarried block in over a millennium. It was moved to Florence, slowly, by oxen and men over a muddy and tremulous two years and sat for another thirty years, undisturbed, except by weather and birds.

We, too, are a kind of sensory detritus, left over after selection down from all possible motor and sensory input. This basic trick—a message and its copy—writ large, is how many creatures of our world create a feeling of the subjective inside. From it, we each create, in all our waking moments, our own marble masterwork. Consciousness is neither sculptor nor sculpture but both, all at once. All brains of sufficient complexity are constantly silencing their own expected movements, and the body outline of the self is actively created by these processes, by making bits of the future and stepping into it like a superhero does their suit. Anna's subjective world is simply what is carved out of the difference between her brain's prediction of where she should be relative to where she is.

A Small (or Large) Learning Machine Made Out of Words

I n a scene from Leo Tolstoy's *Anna Karenina*, a character, Kitty, imagines the meeting of two men—an unrequited crush and a lover—at a party that she is about to attend. She imagines how the men, who have never themselves met, might interact, with Tolstoy writing that she "endlessly considered in her mind, first each of them separately, then the two together."

Such conscious simulations are common, useful, and one of the key learning abilities separating natural from machine learning.[1] Some such simulations are conscious and intentional; others are conscious and spontaneous, and seem to appear by fiat. What the psychologist William James called the "flights" and "perchings" of consciousness, what others have called a "stream," and what some think "sits stiller than a tree" are not just idle or inert imaginings. They guide, dictate, and make more efficient the process by which experience shapes atoms so that those atoms can more efficiently guide experience. They help the human brain generalize similar features of objects, locations, or people without having to experience them in every possible situation. They act as a kind of educated practice and, as data from the results of the simulations accumulate in the mind, the results can be used for ever further simulations, like a Lego brick taken from its original set to use in infinite others. Eventually, the data is cut and carved enough that new data sets can be generalized from the abstractions and these generalizations gather into something that can seem, over time, like reasoning.

An organism's immediate environment can constrain such thoughts, as bumpers do a bowling ball or a gorge does a river. When the Stoics engaged in what is sometimes called "the premeditation"—or the *futurorum malorum premeditatio*, Latin for "prestudying bad future"—they were learning. When samurai warriors visualized the worst thing that could happen in battle ("One who is supposed to be a warrior considers it his foremost concern to keep death in mind at all times, every day and every night") to steel them against horror, they were learning.[2] What Rudyard Kipling said about a day's military march haunting the night, that there is "no discharge in the war," or what the French call *l'esprit de l'escalier*, or "staircase wit," are all conscious simulations.[3] These are not bugs in the conscious system. They are features.

In an old *Twilight Zone* episode, titled "A Penny for Your Thoughts," a man is granted the temporary ability to hear other people's thoughts. The man goes into his work at a bank and finds the elderly security guard at back, whose thoughts surprise him—the guard is rehearsing a detailed plan to rob the bank and escape to Bermuda. Of course, the guard was just bored. He imagined robbing the bank every day, mostly just to pass the time.[4]

I immediately recognized the old guard's ways. As a student, I worked in an art museum, and used to imagine all the ways I would steal the art. I observed the long hallways with the foveal security cameras pointed only down and the habits of the guards in the security booth, who were, at the end of the day, distractable. And then, during graduate school, slightly less feloniously, I used to replay a scenario in my mind over and over where one of my lab's mind-controlled, parasite-infected mice escaped and I had to catch its tail with my bare hands lest I start, or at least encourage through Seinfeldian inaction, the apocalypse.

This turned out to be useful: once, a mouse jumped its maze, in an incredibly dark room with only a few infrared lights shining. Near instantly, during the mouse's midsprint, I picked it up by the tail. In the samurai sense, had I already learned how to, without ever having to actually do so? The mind doesn't jump in a random walk to random thoughts—it flows as a pattern completion, and just because thoughts seem random does not mean they are. Like the character from *Anna Karenina*, I had been running relevant, conscious simulations in the halls where I would likely spend my days and nights—I did not imagine trying to catch the mouse in a Grecian hippodrome or on the moon, after all; and Kitty did not imagine the unrequited love and the lover meeting in Berlin, or underwater, or in a Grecian hippodrome, or on the moon, after all.

We spend almost half of our waking thoughts reliving memories or planning for the future.[5] The challenge for understanding how such thoughts guide learning is that capturing any thought is difficult, because the act of reporting a thought, like the act of locating any single electron, also changes it. And then of course there is the further question, common to anglers, of what to do with them once caught.

Each neuron in the brain, of which there are more than fifty billion, has two genomes, countless proteins, fragments of stray genetic code, and up to tens of thousands of synaptic connections to other cells, each with their own physical and electric stores. And even these estimates are, likely, dramatic underestimates of the brain's potential parameters, or variables, useful for statistical learning. Every brain has vastly more stores than all modern AIs and machines combined. Biology is messy at the level of its atomic and molecular happenings, but contained in that messiness is a staggering amount of ways to be.

Consciousness evolved because it is an incomparable playground for learning. Mind wandering is not random. It is purposeful, careful, and efficient. Natural learning can bring, with apparent ease, the totality of prior experience to bear on any new problems faced, which is not true for modern AI systems, which can be worse than useless at tasks they were not trained on. Modern autonomous car companies, for example, train their artificial drivers virtually, on simulated roads. One advantage to this approach is that the actions of the "virtual fleet" can be sped up considerably and the lessons from each car's experiences can be combined. What one car learns, they all do.[6]

Some claim that their cars drive more than twenty million simulated miles, or the equivalent to more than one hundred years in the real world, every day.[7] The cars are powered by machine-learning algorithms that rely heavily on learning by experience, either simulated or real; thus, it is seemingly natural to describe "simulation" time in comparison to the "actual" time a human would have taken, if only to give an impressive sense of scale.

Such comparisons, however, are following what has become a kind of standard, or at least an increasingly common, kind of explanation in the machine-learning field. One company said an AI robot had played "180 years' worth of games against itself every day" and that, in total, it had been practicing the equivalent of 24–7 for around forty thousand years, or about as long as humans have been painting caves.[8] Another wrote that one of their AIs had relied on "the equivalent of thousands of hours" of simulated game playing to be as good as a person at certain video games.[9]

Why can't the expert AIs just pick up a new game and learn it, as we do, in a matter of minutes? An explanation for this difference between human and artificial learning is one of the thorniest problems in AI research; to solve it, in some sense, would be to also solve the key piece on the road to artificial general intelligence, or AGI, a hypothetical AI that would be as good at all tasks as humans are, including speech, action, art, thought—anything.

What sounds like the computer's large simulation training times are necessary, in part, because their learning often has to re-create relevant parts of natural evolution that we are born with. An awful lot of the massive amounts of simulated hours of training that these artificial systems require is to pull them up to the level that evolution started us from.

Thus, in order to truly compare AI learning with natural learning, there needs to be a better understanding of how to clarify, count, and define what happens as a human brain learns. What is an "hour" of human thought? One would need to understand the full scale of simulation that the human brain performs, both unconsciously and consciously. And, if the unconscious thoughts guide the conscious ones in some way, which they almost certainly do, the human total would need to be uncompressed and summated in parallel across each of its simulations, as well. There is no reason to believe that small parts of our brains are not themselves conscious but their ruminations merely inaccessible—dreaming during the day, say, of all the things that did once or soon might happen.

The brain's dark matter, the unconscious, here-be-dragons part of its activity map, is not just coordinating gesture and language but also simulates via prediction and helps us arrive at solutions to puzzles or dilemmas that appear to have arrived by fiat, muse, or genius. But they did not. They arrived as the conscious mind simulated itself and rued the ways it might have been. It learned on its own simulations. Like the cars, these simulations are sped up and parallelized: a mouse, seconds after it is plucked from a maze,[10] shows signs of neural activity as if it is replaying, at a rapid speed, and sometimes backward, pieces of the maze it just left; people with amnesia who played a *Tetris*-like game, though they did not remember doing so, all became better at the game and some reported seeing, while awake, strange, geometrically shaped "images that are turned on their side" that were "like blocks," but could not explain why; the visual cortex of people who played the same game for the first time showed reactivation of some of the same patterns, at high speed, of each tetromino during the beginning of sleep.[11]

Though the total number of hours the human brain spends simulating a problem unconsciously is difficult to estimate, it is likely vast and can happen independent of memory or awareness. The brain seems to be constantly rehearsing all that it has recently done and might soon do. One "hour" of a person driving or playing a game does not end when the task does; similar to the cars, we are simulating, speeding up, and replaying experiences all day, every day.

I once played a video game that hides puzzles in objects' shadows. Out in the real world, my brain started solving illusory "puzzles" in real shadows, too.[12] My mind had cued in, after only a few hours of playing a game, to the idea that shadows were more relevant than they ever had been, even though other parts of my brain knew very well that I wasn't still playing the puzzle game. This phenomenon happens below conscious awareness, in the brain's attempt to enhance sensory prediction and simulation of a confusing, ever-changing outside world, which implies that perception itself is also a kind of simulation. Should we count these, too, as simulation hours or thoughts?

We are constantly predicting at different timescales into the future, from the millisecond saccades of eye movements to the larger, conscious simulations, much closer to what we might colloquially refer to as "thinking." These allow us to imagine and decide about events that have not occurred, but might: a sailor, for example, deciding to avoid storm clouds (hours); whether to take antibiotics (days); whether to invest in a 401(k) (decades); what to do about nuclear waste (millennia); or while reading Robert Frost's "Fire and Ice" (eons).

While it is true that we primates need only a few minutes to learn the guiding principles of a new task, the tasks often repeat existing human skills, like seeing, moving, speaking, coordination of the hands, or modeling other people's minds. The timing of those few minutes matters greatly. A human infant a few minutes after birth is useless at all tasks except suckling and crying, while the robots, given little guidance, start more like infants, but only if that infant also had to learn the rules of the game or task alongside all the rules of perception, action, and the statistics of their universe.

Human infants have major learning advantages over robots as they age because they do not have to learn how to learn efficiently but come, in a sense, preprogrammed with all the rules needed to grow from a single cell into a denuded, smartly learning primate. You could call your lifetime of experience your age—or you could call it your age plus three billion years.[13]

Once, in my luggage, en route to a neuroanatomy lecture I was leading, I had in front of me a monkey brain, a human brain, a deep-brain stimulator, and a computer's hard drive. Should what the primate brains once were dictate what they are now? Because the human brain was / is / used to be / used to contain / emerged as / grew into a middle-aged woman, from North Dakota, who had died a few years earlier, and the monkey brain was / is / used to be / used to contain / emerged as / grew into a laboratory monkey. The others were rote: a computer that once used electricity to control a brain and a physical copy of a computer's memory, inert and cold.

The most obvious similarity between the items is that all of them were off—powered-down versions of what they once were. Each of the four, in a sense, is also a kind of information processing and storage technology, yet only two of them were once conscious. Two biological hunks of matter, both previously conscious, and two artificial hunks of matter, neither of which was conscious. Yet all four items could, at one point, compute and sense things, and all four used electricity as their commands when "on." And it gets even stranger. The deep-brain stimulator, which is basically a pacemaker with its electrodes turned upward into the brain, was like a bridge between the natural and artificial, for it could impinge on neurons and change consciousness by shaping and shunting electrons around, just like the hard drive, just like parts of the monkey and human brain could to its other parts. The stimulator could, if tuned correctly and if placed in the right spot in a brain, enliven a comatose person with its electricity, induce laughter and merriment, or suppress a seizure, like an automatic sprinkler might.

Another clear difference is how the data is stored. The hard drive, even while off, retains its information well enough (and perhaps even perfectly), so when or if it is ever rebooted, all the information is still in there. This is also true for the deep-brain stimulator, which has both a structure and function and a kind of memory as firmware or software. Not so lucky with the monkey and human brains, though, which were both preserved with formaldehyde, which hardens proteins into rigid, nonre-bootable lattice structures. But where did the information that the brains once held go? When the monkey's last neuron had kicked out its final discharge and collapsed for good, like a burst balloon, what happened to the lifetime of its monkey skills and monkey memories? Were they still in there, somewhere?

Information in computers is stored as limits on the possibilities of where electrons can go. A one-terabyte hard drive full of data weighs approximately 10^{-15} grams more than an empty one and a full, four-gigabyte e-book weighs a billionth of a billionth of a gram more than an empty one, but a monkey brain full of memories weighs no more than an empty one and a conscious monkey brain no more than an unconscious one.

Modern training of artificial intelligences, dominated by machine learning, has made the hard drive more like natural brains in the sense that the machines now act on probability and expectations, which are mostly based on prior observation and action. But what if, on the hard drive, I had kept one of those fancy new language robots, or Go-playing robots, or the algorithms to a self-driving car? Is there any sense in which the stored data could be considered, like the preserved brains, a reflection of what once was / is / used to be / used to contain / emerged as / grew into?

The biggest difference is that the machines were never conscious and so could not learn properly enough to become conscious themselves, which is why it is

harder to create a robotic soccer fan than a robotic soccer player, harder to create a sex-enjoying robot than a
sex robot, harder to create a robot self-conscious about
its lack of wit than a witty robot. Today's machines,
even the fancy learning ones, will never be conscious
for the same reason a computerized weather simulator will never be "wet"—or, maybe, they just haven't
learned enough yet.[14] Perhaps the creation of a truly
general learning robot will require, as natural selection
did for us, the ability not just to learn but to consciously
rue a past and daydream of a future, near and far, filled
with more than just electric sheep.

Not That Hard

*The principles of simplicity, elegance, and even beauty
that drive physicists' search for a fundamental theory
will also apply to a theory of consciousness.*

—**David Chalmers**

For thousands of years, people thought the sun revolved around the earth because it looked like it did. The right question to ask, however, was, What would the sun's arc in the sky be if the opposite of what it merely looked like were true? The answer, of course, is that it would appear identical and that naked-eye observation, induction, and intuition alone could not distinguish which of the competing ideas was correct.

There is a difference between theories of how a brain works and theories of consciousness. The unsolved problems of consciousness can be categorized into many easy problems and one hard one. The easy problems deal with how the brain works. The hard problem, and there is only one, is the need to explain why it feels like anything at all to be conscious.[1]

When, in the late twentieth century, a girl undergoing brain surgery for epilepsy laughed after the surgeon stimulated the part of her brain that connects to the muscles in her face and throat, an explanation for the causal chain of events from the surgeon's wand all the way down to the mechanical reason her muscles fired is an easy problem. Those studying the brain will likely be able to explain these events, mostly, and in the sequence they happen, with only a few more centuries of good work.

A full explanation for how such activity led to / was coincident with her subjective feeling of mirth and joy is the hard problem—which may not be solved, ever.

There are many ways to understand the stakes of the hard problem, but the central questions are clear enough: Why should consciousness feel like anything at all? Why, when an adult woman wakes up every morning, does it feel like she looks like something rather than nothing? Why, while asleep, does it feel like a version of her is living through something conscious even if the rules are all fun house? Yet nothing about any theory of how the brain works tells us definitively how the feeling of subjectivity unfolds from its workings. Broadly, there are only two possibilities: either a full description of the brain's workings will yield an explanation for the feeling of being conscious or a full description of how a brain works will always be inadequate and cannot yield such an explanation.

There are many small mysteries of consciousness that blur the easy and the hard problems and that, to this day, remain unexplained: most people cannot tickle themselves, but those with a tendency toward schizophrenia sometimes can;[2] people visually imagine at different subjectively perceived distances in front of them; if a human brain is cut in half along the cables connecting the left and right hemispheres, two consciousnesses emerge, and yet twin girls who share a physical connection between a deep structure in their brain that coordinates sensory information claim to be able to guide each other's attention and feelings;[3] when a person is asleep and dreaming, it is always they, and not other dream characters, who speak, and yet the dreams are only ever from one point of view. These, and a near infinity of other questions, remain unexplained.

The hard problem of consciousness is the scientific study of a blind spot whose borders and terms we cannot yet define. How, from the inside, will we explain what it means to feel like being from the inside? Consider a thought experiment where we imagine a primate species, in almost every way identical to us except it evolved without eyes or vision of any kind but that, despite this, retained progress toward cities, art, science, mathematics, and civilization. With no eyes to look up and wonder with, we can ask: Would such a species have ever inferred the existence of the moon, and if so how?

Probably, yes. Eventually. But it would take a while.

There would be a long while where *Homo caecus* ("blind man") would notice regularities in nature in tides or other biologic, repeated cycles linked to lunar cycles, but they wouldn't know them as such without the time-based correlation of the nightly lunar arc and change.[4] Without eyes, *H. caecus* wouldn't be able to see the atmosphere, let alone ever look up into the stars and recruit childhood wonder. As a result, astronomers would not exist, as such. *H. caecus* Egyptian, Babylonian, Greek, and Chinese thinkers would leave blank all mention of the heavens above. The world would be an anechoic chamber, not a stage. Only the metaphysically inclined would be interested in astronomy, since religious or explanatory texts wouldn't mention anything visual. The wine-dark sea would be the salt-lick sea.

H. caecus would almost certainly develop echolocation, like a bat, or a kind of self-sensing electroreception, similar to the electric fish, and the concept of horizon wouldn't apply to morality, empathy, or goal-directed behaviors. Explorers would turn in circles. Manifest destiny would end at the nearest canyon wall. The moon, if imagined or described, would sound impossibly high in the sky, like descriptions of clouds do to the congenitally blind *H. sapiens*. The position of the sun, at least, could be inferred by its heat on one's cheek or, maybe, with rates and locations of skin cancers.

And because nobody would have passed on the desire to their children to pursue physics or astronomy,

Galileo, Newton, and Einstein would bide their time echolocating other pursuits. Some things may have arrived sooner. A prototheory of gravity, perhaps, would have been worked out thousands of years earlier, in Babylon, Greece, or China, as sonic and tactile interactions took intellectual and spiritual precedence over visual ones. But who would dare to claim the gravitational effects of a giant, cold rock that shuts out the heat from the sun? Who in Ancient Greece or China could dare claim that the angry god who covered the sun completely every day was sitting in the sky the whole time and shaping all the oceans?

At some point, someone would develop an instrument to scan objects using what we would consider the "visual" wavelengths and convert those to sound, a simple process. The instrument would be shaped like a telescope and would be pointed up at the sky by chance, at first, maybe because that is where rain comes from. The instrument would scan the sky one patch at a time and turn any light source into sounds, in the same way that we use tools to detect various electromagnetic frequencies outside our visual range, like infrared or ultraviolet.

And so *H. caecus* would end up with a sound map of the sky, and every time it came across a star it would beep. When it crossed the moon during the night, there would be a huge, unexplained acoustic blob the approximate size of the sun up above them, which became crescent and full and waxed and waned on an unexplained cycle. Those who attempt to explain this would try their best but get it wrong for a while.

And then at some point there would be a theory and that theory would require the observation from ever larger acoustical telescopes that could map the entire sky at once. A lot of effort would go into making the acoustic equivalent of a visual map, which has the advantage of simultaneity, and so the solution would be, probably, orchestral. Scientists would be composers of symphonies and vice versa; great thinkers would explain the tides and other natural, circadian, or biological periodicities using sonic terms of art, but all of them would be wrong in a way, for none would yet have spotted the influence or existence of the strange, mute rock hovering above them.

Eventually, however, it would be natural to leave the planet, at which point orbital mechanics and geotectonic stress would come into play. Any sort of GPS would require the mathematics of the two-body problem and the moon would be inferred, as a radiating force, in the equations. There would have been hints, in retrospect, all over the place. Still, no *H. caecus* would have ever seen the moon. There would be no "face" in the moon, but perhaps those inclined to false-positive pattern recognition would find in its song a kind of hidden, illusory "chirp." The first astronaut to step on the moon would take a small piece of its sand and taste it.

We are similarly blind trying to investigate consciousness from within. It is a blind search for a cause from its effects. Is there a possible explanation for how we get feelings from small electric storms in our heads? Sure. But are we, *Homo sapiens sapiens,* searching blindly for something we cannot detect with modern instruments? How would we ever know?

Nineteen Ways of Looking at Consciousness

Briefly, putting all the ways together: billions of years ago, in the oceans, molecules exploited natural gradients in hydrogen concentrations in the water, which, like all batteries, like all power storage must be, was simply having more of something on one side versus another ("A Small Town with Too Much Food"). Eventually, it made sense to capture and preserve these gradients—from whence, cells—in order to keep most of the outside out and most of the inside in.

This gave cells the freedom to wander, which meant a way of moving around was useful ("The Anxiety Felt While Prevented from Migrating"), which meant that tracking these movements was useful ("A Sex-Starved Cricket Sculpting in Time"), which meant timekeeping was useful ("The Music While the Music Lasts"), which meant learning what to predict was useful ("A Small [or Large] Learning Machine Made Out of Words"), which meant that finding an efficient way to combine movement, learning, prediction, and sensation was useful ("The Arbiter of Elegance").

Occasionally, by what appears to be chance, but actually isn't, lightning will break a tree in half. A similar process of what appears to be chance, but actually isn't, duplicates genes and clusters of genes and also neurons and clusters of neurons ("The Median Price of a Thrift-Store Bin of Evolutionary Hacks Russian-Dolled into a Watery, Salty Piñata We Call a Head"). These duplications, spread across billions of years, allowed a compressed kind of natural selection that operates not only on individuals or species but also on all cells and all of their junctions and their parts, always. The molecules and clusters of molecules follow the same scaleless laws that the cells do as they grow, which means clusters of cells do, too. Learning loops replicated from basic motor pairings allowed gesture learning, which allowed language learning, and ever more duplications of the same loops allowed the entire conscious apparatus ("Swinging Through Ancient Trees While Standing Still and Hearing Voices").

Thus, the same mechanism that makes internally generated voices sound like they come from inside is the same mechanism that makes all thoughts—memories, imaginings, subjectivity—appear to come from the inside rather than the outside ("A Make-Believe Parasite with No Legs and Places to Go"). All experiences and all thoughts are, broadly conceived, movements that have been suppressed across and narrowed by a kind of terrain, or topography ("Sunlight Raining Down on Gridworld"). Conscious thoughts can be manipulated in simulation, just as speech sounds can, and just as the hand can, but only once they are selectively broadcast, with constraint, throughout the brain to the large swaths that want to know ("An Ante Meridiem Radio Drama").

The feeling of what it is like to be something is necessary for learning about learning and is different from unconscious gesture learning, because the feeling is carved out of the cells that respond to the outside world's irritations ("Not That Hard"). What we call "thinking," thus, is manipulation of practice gestures, where gestures are thoughts derived from learning loops and conscious thoughts can be manipulated as inputs to a radio broadcast, a network of causes and effects, a collapse of quantum uncertainty, or a lie. The conscious illusion—it is an illusion, after all ("Relative to the Observer Who Is Also a Liar")—being either a kind of analog qua digital blended holography ("Like the Rise and Fall of Pinball") or a kind of simulated, virtual reality ("A Simulation Starring You"), which may draw its power to defeat determinism from the small, crisp spaces in the hollowed-out pieces of cells ("An Itsy-Bitsy Teeny-Weeny Quantum-Dot-like Non-Machiney").

Just as one can move one's arm while also manipulating the fingers—that is, at different scales and levels of movement—one can metalearn, with thoughts about thoughts or learning about learning. A person thinking abstract thoughts in a sensory deprivation tank or who is locked in, or a brain in a jar is doing no different than simulating a hand movement with, instead, memories as its degrees of freedom. In other words, language does not allow consciousness, but the same wiring trick that allows language also allows consciousness. The reason some chunks of tissue feel different from others is because they are constrained by their learning and access, meaning there can be parts of the brain that, if excised, do not alter, or at least only minimally alter, consciousness ("Endeavoring to Grow Wings").

This kind of manipulation of memory endured because thinking is useful, because it allows kinds of learning otherwise impossible. These are the simulations of wandering thought that can briefly be contained, like a firefly, but the boundaries of containment are the selections. The entirety of the "stage play" of conscious experience is the sensation of the outside world but cut off in order to create the boundaries between self and nonself ("A Secondhand Markov Blanket"). Language does not supersede thought, but it, like memory, perception, and all spontaneous thought, is contained within the umbrella of motor movement and discharged as a selection of options pared down until only the one, the chosen one, remains.

Any system is conscious when it can constrain its own degrees of freedom by acting on itself ("The North

African Rhino of Charismatic Megaquale"). The compression is not the consequence but the source. We are
more than pursuers of elegance; we are elegance itself.
"You" are a skillful reduction from options. "You" are
defined by the infinity of ways you are not.

Appendix

Electric Current Stimulates Laughter

Note: This paper is reproduced, with permission from the author, without figure 1, a visual depiction of the stimulated brain sites. The figure can be found in the original paper.

Source: Fried, I., Wilson, C., MacDonald, K. et al. Electric current stimulates laughter. *Nature* 391, 650 (1998).

Abstract

Speech and laughter are uniquely human. Although there is considerable information on the neuronal representation of speech, little is known about brain mechanisms of laughter. Here we report that electrical stimulation in the anterior part of the human supplementary motor area (SMA) can elicit laughter. This area is also involved in the initiation of speech and has been shown to have increased activity in people who stutter.[1]

Main

Electrical stimulation was applied at 85 discrete sites on the cortical surface of the left frontal lobe of a 16-year-old girl (A.K.) undergoing monitoring by intracranial subdural electrodes to locate the focus of chronic intractable seizures. The patient's seizures were never accompanied by laughter. During stimulation A.K. performed a variety of tasks including naming of objects, reading a paragraph of text, counting, rapid alternating supination and pronation of the forearms, finger-to-thumb apposition, and alternating flexion and dorsiflexion of the feet.

A small area measuring about 2 cm x 2 cm was identified on the left superior frontal gyrus where stimulation consistently produced laughter. The laughter was accompanied by a sensation of merriment or mirth. Although it was evoked by stimulation on several trials, a different explanation for it was offered by the patient each time, attributing the laughter to whatever external stimulus was present. Thus, laughter was attributed to the particular object seen during naming ("the horse is funny"), to the particular content of a paragraph during reading, or to persons present in the room while the patient performed a finger apposition task ("you guys are just so funny . . . standing around").

The duration and intensity of laughter increased with the level of stimulation current. At low currents only a smile was present, while at higher currents a robust contagious laughter was induced. It was also accompanied by the cessation of all activities involving speech or hand movements. At neighboring sites, speech arrest was provoked without the evocation of laughter, and at medially adjacent sites stimulation arrested all manual and speech activities. These sites were anterior to the area where electrical stimulation evoked complex responses involving lower and upper extremities typical of the supplementary motor area.[2,3] The area encompassing sites of laughter, speech arrest, and manual activity extended to the dorsal convexity of the hemisphere, in agreement with current notions about the lateral border of the SMA.[4]

Pathological laughter has been described in several clinical conditions, including pseudobulbar palsy and gelastic seizures associated with lesions in the hypothalamus or temporal lobe.[5,6] However, laughter in these conditions is rarely associated with the appropriate emotional experience, and often the condition of the patient (for example, a seizure) precludes reporting of the emotional experience. There are very few reports of laughter evoked by electrical stimulation at cortical sites including the anterior cingulate and orbitofrontal cortex[7] and the basal temporal lobe.[5] Arroyo *et al.* postulated dissociation of the motor program of laughter and the experience of merriment, localizing the former in the anterior cingulate and the latter in the temporal lobe;[5] however, our data show that laughter and mirth can be evoked by stimulation of the frontal cortex in the anterior part of the SMA, an area associated with the execution of motor programs.

The observation that A.K. was able each time to invoke a stimulus context that "explained" the laughter suggests a close link between the motor, affective, and cognitive components of laughter. It is likely that these varied components are represented in a large neuronal network capable of parallel distributed processing, where the entire network is activated as a whole by the stimulation of any of its constituent units.[8] Our observations also suggest that smiling and laughter might involve similar mechanisms and are closely related phenomena on a single continuum.

Although it could be argued that the neural representation of laughter observed in our epilepsy patient might not reflect the substrate for laughter in the normal brain, it should be emphasized that laughter was not a part of A.K.'s seizures. Furthermore, the region of seizure onset was medial to the area of evoked laughter, and stimulation in this region did not produce laughter responses. The human SMA has somatotopic organization, with the hands represented anterior to the arms and legs, and face and speech represented most anteriorly.[3] Our results suggest that speech and laughter are closely represented in the rostral part of the SMA, just anterior to the representation of manual activity.

We propose that the anterior part of the SMA is part of a further development in humans to accommodate the specialized functions of speech, manual dexterity, and laughter. This area might correspond to the presupplementary motor area, a region situated anterior to the SMA proper, recently described in nonhuman primates, and thought to be involved in high-level motor programming.[9,10]

1. Fox, P. T. *et al. Nature* **382**, 158–161 (1996).
2. Mitz, A. R. & Wise, S. P. *J. Neurosci.* **7**, 1010–1021 (1987).
3. Fried, I. *et al .J. Neurosci.* **11**, 3656–3666 (1991).
4. Wise, S. P. *et al. Adv. Neurol.* **70**, 489–495 (1996).
5. Arroyo, S. *et al. Brain* **116**, 757–780 (1993).
6. Martin, J. P. *Brain* **73**, 453–464 (1950).
7. Sem-Jacobsen, C. W. in *Depth-electrographic Stimulation of the Human Brain and Behavior* (ed. Gantt, W. H.) 127–138 (Charles C. Thomas, Springfield, Illinois, 1968).
8. Rumelhart, D. E., & McClelland, J. L. *Parallel Distributed Processing: Explorations in the Microstructure of Cognition* vol. 1 (MIT Press, Cambridge, Massachusetts, 1978).
9. Tanji, J. *Curr. Opin. Neurobiol.* **6**, 782–787 (1996).
10. Picard, N. & Strick, P. *Cerebral Cortex* **6**, 342–353 (1996).

Acknowledgments

I would like to thank Robert Sapolsky and Christof Koch, for intellectual roost; my indomitable agent, Lauren Sharp, for the iterative flight plan; editors Pronoy Sarkar and James Meader for the flight lessons; George Witte for the branch; Julia Icenogle for wing repair; Ken Martin for ailerons; Dr. Matthew MacDougall, Peretz Partensky, and Cam Christie for birdseed; Aaron Bornstein and Brian Laidlaw for aerial drills; and Marie House, of course, for the hatching.

For support, I would also like to thank the Allen Institute for Brain Science, which hosted me as writer-in-residence.

I interviewed dozens of people over the course of researching this book and the article on which it is loosely based, "What Is Elegance in Science?," which was published online, in 2015, by *The New Yorker*. Each conversation shaped my thinking and this manuscript in ways that are impossible to fully describe. I would like to especially thank, for their time: Lisa Randall, Ed Witten, Alexander Wissner-Gross, Sydney Brenner, Edvard Moser, Heather Knight, Winfred Menninghaus, Stephen Smith, Diana Tamir, Blaise Agüera y Arcas, Jonathan Blow, Dr. Jonathan Leong, Andy Clark, Dr. Adam Zemen, Iké

Udé, David Poeppel, Eric Haseltine, Alexander Nemerov, Edward Tufte, Alyssa Goodman, Jonathan Corum, and Dr. Joel Zivot.

This book would not be possible without the contributions of authors and scientists who came before me. Specifically, this book relies heavily on the ideas and work of those mentioned above and also: Dr. Itzhak Fried, Nick Lane, Daniel Dennett, Dr. Giulio Tononi, Dr. Karl Friston, Alva Noë, Peter Godfrey-Smith, Dr. Rodolfo Llinás, A. G. Cairns-Smith, Francis Crick, Douglas Hofstadter, Peter Todd, and David Chalmers.

I, and we all, owe much to their work.

Notes

· · · · · ·

Chapter 1: Relative to the Observer Who Is Also a Liar

1. Pain receptors, or nociceptors, detect the many kinds of stimuli a body might experience as painful. However, these receptors are not found in the central nervous system itself, which receives messages about peripheral pain but cannot generate them.

2. The book is 3.19 pounds; the human brain, on average, 3.3 pounds.

3. The ship of Theseus stars in a classic thought experiment where wooden planks are replaced, one at a time, until the whole ship is eventually turned over. Whether the ship remains the same after these modifications has been discussed by philosophers for centuries, since at least ancient Greece, with metaphorical applications to identity and selfhood.

4. Descriptions of the effects of anesthesia were drawn from a phone conversation with Dr. Joel B. Zivot, an anesthesiologist, and from: George Alexander Mashour and Ralph Lydic, *Neuroscientific Foundations of Anesthesiology* (New York: Oxford University Press, 2011).

5. There is little better within-person control condition for the empirical study of consciousness than the remarkable fact that, some nights, during sleep, we have experiences so vivid and convincing that we can rarely, if ever, tell the difference between them and waking life. It remains a complete and total mystery why we sleep and, further, why we dream: Gian Gastone Mascetti, "Unihemispheric Sleep and Asymmetrical Sleep: Behavioral, Neurophysiological, and Functional Perspectives," *Nature and Science of Sleep* 8 (July 12, 2016): 221–38; Patrick McNamara, *The Neuroscience of Sleep and Dreams*, Cambridge Fundamentals of Neuroscience in Psychology (Cambridge: Cambridge University Press, 2019).

6. Scientists have wondered whether these conscious dream

experiences, just as real as any other, are more like imagination or perception. To study this, they have devised methods for communicating with people in dream states: Benjamin Baird, Stephen LaBerge, and Giulio Tononi, "Two-Way Communication in Lucid REM Sleep Dreaming," *Trends in Cognitive Sciences* 25, no. 6 (June 1, 2021): 427–28; Stephen LaBerge, Benjamin Baird, and Philip G. Zimbardo, "Smooth Tracking of Visual Targets Distinguishes Lucid REM Sleep Dreaming and Waking Perception from Imagination," *Nature Communications* 9, no. 1 (August 17, 2018): 3298; Sergio Arthuro Mota-Rolim, "On Moving the Eyes to Flag Lucid Dreaming," *Frontiers in Neuroscience* 14 (April 15, 2020): 361.

7. There is something like a forty-million-year gap in the fossil record, but cetaceans, which include whales and dolphins, and hippos are each other's closest living relatives. For one take on the debate on their genomic similarities: Mark S. Springer et al., "Genomic and Anatomical Comparisons of Skin Support Independent Adaptation to Life in Water by Cetaceans and Hippos," *Current Biology* 31, no. 10 (May 24, 2021): 2124–2139.e3.

8. The oceans have a "deep sound channel," a horizontal layer that, because of its pressure and temperature, allows low-frequency sounds to travel thousands of miles with little dissipation.

9. For more on what those who get their vision restored perceive when they first open their eyes: Richard Held, Yuri Ostrovsky, Beatrice de Gelder, Tapan Gandhi, Suma Ganesh, Umang Mathur, and Pawan Sinha, "The Newly Sighted Fail to Match Seen with Felt," *Nature Neuroscience* 14, no. 5 (May 2011): 551–53; Patrick House, "What People Cured of Blindness See," *New Yorker*, August 28, 2014.

Chapter 2: Like the Rise and Fall of Pinball

1. The neuroscientist Christof Koch, in conversation.

2. The modern brain, like the modern pinball machine, has both electrical and mechanical parts; it sends and receives both analog and digital information. Neither started out that way. Of the two histories, the evolution of pinball machines is better documented, because we were there for it.

3. A line from arguably the finest pinball-themed essay of all

time: John McPhee, "The Pinball Philosophy," *New Yorker*, June 30, 1975.

4. Details of bagatelle, the early electrification of the game that became pinball, and the commercial and technical struggles in the twentieth century can be found in Greg Maletic, director, *Tilt: The Battle to Save Pinball*, Solid Entertainment, 2008; "The Bagatelle Wizard Instead of the Pinball Wizard," National Museum of American History, October 31, 2012; Michael Shalhoub, *The Pinball Compendium: Electro-Mechanical Era*, ill. ed. (Atglen, Pa.: Schiffer Publishing, 2008).

5. A classic study, showing how readily we attribute narrative to chance interaction or abstracted behavior: Fritz Heider and Marianne Simmel, "An Experimental Study of Apparent Behavior," *American Journal of Psychology* 57, no. 2 (1944): 243–59.

6. Though the brain is not a computer, it is computational, by some definitions. More on the analog and digital peculiarities of cells in the human brain: Dennis Bray, *Wetware: A Computer in Every Living Cell*, ill. ed. (New Haven: Yale University Press, 2011); Donald S. Faber and Alberto E. Pereda, "Two Forms of Electrical Transmission Between Neurons," *Frontiers in Molecular Neuroscience* 11 (November 21, 2018): 427; Yasuhiro Mochizuki and Shigeru Shinomoto, "Analog and Digital Codes in the Brain," *Physical Review E* 89, no. 2 (February 4, 2014): 022705.

7. For summaries on the evolution of the brain's disparate but parallel chemical and electrical communication systems: William B. Kristan, "Early Evolution of Neurons," *Current Biology* 26, no. 20 (October 2016): R949–54; Pedro Martinez and Simon G. Sprecher, "Of Circuits and Brains: The Origin and Diversification of Neural Architectures," *Frontiers in Ecology and Evolution* 8 (2020): 82; Anthony N. van den Pol, "Neuropeptide Transmission in Brain Circuits," *Neuron* 76, no. 1 (October 4, 2012): 98–115.

Chapter 3: The Anxiety Felt While Prevented from Migrating

1. This chapter draws heavily from the work of neuroscientist Rodolfo Llinás, who has argued that the control of movement is the central force shaping the evolution of the brain,

going so far as to claim that the brain's control of organized movement "gave birth to the generation and nature of the mind": Rodolfo R. Llinás, *I of the Vortex: From Neurons to Self*, repr. ed. (Cambridge, Mass.: Bradford Books, 2002) and "Intrinsic Electrical Properties of Mammalian Neurons and CNS Function: A Historical Perspective," *Frontiers in Cellular Neuroscience* 8 (2014): 320.

2. Llinás once speculated, based on observations of a so-called "physiological tremor," where the muscles spasm rhythmically at around ten times a second, that such a tremor is endemic in all of us and serves as a kind of metronome for the timing of all movement: R. Llinás, K. Walton, D. E. Hillman, and C. Sotelo, "Inferior Olive: Its Role in Motor Learning," *Science* 190, no. 4220 (December 19, 1975): 1230–31; R. Llinás and J. P. Welsh, "On the Cerebellum and Motor Learning," *Current Opinion in Neurobiology* 3, no. 6 (December 1993): 958–65.

3. This phrase is from the poet Robert Pinksy, who wrote about the physicality and movement behind learning poetic sound: "The hearing-knowledge we bring to a line of poetry is a knowledge of patterns in speech we have known to hear since we were infants . . . And yet, having learned these graceful, peculiar codes from the cradle. . . .": Robert Pinsky, *The Sounds of Poetry: A Brief Guide* (New York: Farrar, Straus and Giroux, 1999).

4. In mechanics, an object's degrees of freedom are the number of parameters required to describe its position. The human hand, influenced by the shoulder, elbow, and wrist joints, has at least twenty-five degrees of freedom, a remarkable number probably unparalleled in the animal kingdom. The Soviet neurophysiologist Nikolai Bernstein first described the paradox of the "degrees of freedom problem" by arguing that the near-infinite ways an animal can move should present more of a problem to the brain than it appears to: N. Bernstein, *The Co-ordination and Regulation of Movements*, (Oxford: Pergamon Press, 1967); N. Bernstein, *Dexterity and Its Development*, ed. Mark L. Latash and Michael T. Turvey (New York: Psychology Press, 2015); Hiske van Duinen and Simon C. Gandevia, "Constraints for Control of the Human Hand," *Journal of Physiology* 589, no. Pt 23 (December 1,

2011): 5583–93; Raffi Khatchadourian, "How to Control a Machine with Your Brain," *New Yorker*, November 19, 2018.

5. I'll be honest. I stole this idea from the opening scene of the 1988 film *Dead Ringers*.

6. One implication of limits to the pacing of movement would be, speculatively, that the temporal limits of movement and thought are linked somehow.

Chapter 4: The Music While the Music Lasts

1. The neuroscientist Antonio Damasio used this phrase to describe consciousness and the self, arguing that the "I" is neither the creator of nor listener to the stories inside our brains—it is, instead, the stories themselves, so long as they keep playing: Antonio Damasio, *The Feeling of What Happens: Body and Emotion in the Making of Consciousness* (San Diego: Mariner Books, 2000).

2. One of the earliest theories of consciousness to have testable hypotheses grounded in empirical data involved patterns of coordinated electrical activity across the brain, called oscillations. One such theory was organized by the geneticist Francis Crick and the neuroscientist Christof Koch in a series of collaborative scientific papers. Changes in consciousness brought on by sleep, anesthesia, focus, and injury appear alongside changes in brain oscillations, meaning the relationship between the two is at least coincident, if not causal: G. Buzsáki, *Rhythms of the Brain* (Oxford: Oxford University Press, 2006); Francis Crick, *The Astonishing Hypothesis: The Scientific Search for the Soul* (London: Scribner, 1995); Francis Crick and Christof Koch, "Towards a Neurobiological Theory of Consciousness," *Seminars in the Neurosciences* 2 (1990): 263–75.

3. I don't mean this literally. I got the fanciful idea from the novelist Julio Cortázar, who once wrote, "When you go to Europe, the soul takes about three days longer to get there."

4. Pink noise, otherwise known as 1/f noise, appears often as a phenomenon in the statistics of biology. For lay and technical reading on the basics of noise, its prevalence in biological systems, and why pink noise has a kind of statistical memory, see: J. M. Halley, "Ecology, Evolution and 1/f-Noise," *Trends in Ecology and Evolution* 11, no. 1 (1996): 33–37; T. Musha

and M. Yamamoto, "1/f Fluctuations in Biological Systems," in *Proceedings of the 19th Annual International Conference of the IEEE Engineering in Medicine and Biology Society: "Magnificent Milestones and Emerging Opportunities in Medical Engineering,"* 6: 2692–97, 1997; Meghan Neal, "The Many Colors of Sound," *Atlantic*, February 16, 2016.

5. Fascinating recent work on a "consciousness meter" uses a statistical measure of complexity of a brain's response to a magnetic perturbation to give a rough measure of how conscious an individual is: C. Koch, "How to Make a Consciousness Meter," *Scientific American* (November 2017): 28–33; Fabio Ferrarelli et al., "Breakdown in Cortical Effective Connectivity During Midazolam-Induced Loss of Consciousness," *Proceedings of the National Academy of Sciences of the United States of America* 107, no. 6 (February 9, 2010): 2681–86; G. Tononi, "Consciousness and Complexity," *Science* 282, no. 5395 (December 4, 1998): 1846–51.

6. The idea that different states of consciousness amount to a phase change, with analog to the chemistry and physics of water during transitions from solid to liquid to gas, has precedent: Max Tegmark, "Consciousness as a State of Matter," *Chaos, Solitons and Fractals* 76 (July 2015): 238–70.

Chapter 5: A Secondhand Markov Blanket

1. The psychologist William James once wrote, "Take a sentence of a dozen words, and take twelve men and tell to each one word. Then stand the men in a row or jam them in a bunch, and let each think of his word as intently as he will; nowhere will there be a consciousness of the whole sentence." This idea, more broadly—defining, or separating, an emergent or larger whole from the boundaries of its parts—is sometimes referred to as the "superposition problem," based on a concept borrowed from quantum mechanics that allows quantum states to be additive ("superposed") and thus allows every quantum state to be described as the sum of two or more others: William James, *The Principles of Psychology: In Two Volumes*, vol. 1, 1890, facsim. ed. (New York: Dover, 1995).

2. Human ears have tens of thousands of tiny hair cells that move, like grass in the wind, in response to the physical

compression of air, which is then interpreted by the brain into what we call "sound."

3. This does happen. The single-celled parasite *Toxoplasma gondii*, for example, can be passed from mother to fetus and hides, often dormant, inside of neurons in the brain for what may be the lifetime of the host.

4. This example is used by the neuroscientist Karl Friston to differentiate, in part, life from not-life. The basic idea is that cellular life, unlike a drop of oil, maintains its borders instead of diffusing, with interesting consequences: Karl Friston, "Am I Self-Conscious? (Or Does Self-Organization Entail Self-Consciousness?)," *Frontiers in Psychology* 9 (2018): 579 and "A Free Energy Principle for Biological Systems," *Entropy* 14, no. 11 (November 2012): 2100–21.

5. Friston also argues that one of the consequences of life's tendency to pursue the minimization of "free energy," a concept related to entropy and equilibrium, involves a way of defining a statistical border between self and not-self. These borders, if defined or arranged properly, can group into something called a Markov blanket, a concept used commonly in machine learning, and possibly guide the organization and behaviors of all life: Karl Friston, "Life as We Know It," *Journal of the Royal Society Interface* 10, no. 86 (September 6, 2013): 20130475 and "The Free-Energy Principle: A Unified Brain Theory?" *Nature Reviews Neuroscience* 11, no. 2 (February 2010): 127–38; Michael Kirchhoff, Thomas Parr, Ensor Palacios, Karl Friston, and Julian Kiverstein, "The Markov Blankets of Life: Autonomy, Active Inference and the Free Energy Principle," *Journal of the Royal Society Interface* 15, no. 138 (January 2018): 20170792.

6. In my vote, one of the most compelling and as-yet-unanswered scientific questions, second only to those related to the origin of consciousness, is whether or not a caterpillar can retain a memory across its goo-stage metamorphosis and on into its moth or butterfly form. The jury is still out: Douglas J. Blackiston, Elena Silva Casey, and Martha R. Weiss, "Retention of Memory through Metamorphosis: Can a Moth Remember What It Learned as a Caterpillar?" *PloS One* 3, no. 3 (March 5, 2008): e1736.

7. In general, monists try to remove boundaries or separations

between concepts. There are many kinds of monism, but I figure if anyone might appreciate a lack of further distinction here, it would be them.

Chapter 6: A Simulation Starring You

1. May-Britt Moser, who won a Nobel Prize for the discovery of grid and place cells, and who lives in Trondheim, Norway, wrote to me after I mentioned my visit to Tyholttårnet: "Yes, I have been there, it is fascinating! The confusion is also that you do not see the rotation from the bathroom, when you come out you use your established map to find your table, but that map has not been rotated since you did not know how fast the outer rim rotated. You have to reset your path integrator by attending to the visual landmarks before you can find your table." (Path integration is a general term for the method animals might use for dead reckoning as they navigate through space.) Moser continued, referencing a study by Fyhn and colleagues, where the researchers induced a version of the Tyholttårnet illusion in rats: "in the supplementary material you will see an example of recordings from hippocampus and entorhinal cortex from two rats in a similar experiment— not rotation, but darkness." One of the rats, she wrote, was nonplussed by the reorientation; the other "needed to be picked out of the environment—briefly—to reset the path integrator." In other words, the illusion might not exist, or exist as strongly, across all individuals for reasons we don't yet understand: Marianne Fyhn, Torkel Hafting, Alessandro Treves, May-Britt Moser, and Edvard I. Moser, "Hippocampal Remapping and Grid Realignment in Entorhinal Cortex," *Nature* 446, no. 7132 (March 2007): 190–94; Rebecca Schwarzlose, *Brainscapes: The Warped, Wondrous Maps Written in Your Brain—And How They Guide You* (Boston: Mariner Books, 2021).
2. May-Britt Moser, David C. Rowland, and Edvard I. Moser, "Place Cells, Grid Cells, and Memory," *Cold Spring Harbor Perspectives in Biology* 7, no. 2 (February 2, 2015): a021808.
3. Edvard I. Moser, May-Britt Moser, and Yasser Roudi, "Network Mechanisms of Grid Cells," *Philosophical Transactions of the Royal Society B: Biological Sciences* 369, no. 1635 (February 5, 2014): 20120511; May-Britt Moser

and Edvard I Moser, "Crystals of the Brain," *EMBO Molecular Medicine* 3, no. 2 (February 2011): 69–71; Hanne Stensola, Tor Stensola, Trygve Solstad, Kristian Frøland, May-Britt Moser, and Edvard I. Moser, "The Entorhinal Grid Map Is Discretized," *Nature* 492, no. 7427 (December 2012): 72–78.

4. Versions of this general idea—that we experience the outside world not as it is, but as our brain intends or needs it to be—have been proposed by philosophers and neuroscientists for centuries, from Plato's "allegory of the cave" to Immanuel Kant's "transcendental idealism" to the philosopher Daniel Dennett's "user illusion" to the cognitive psychologist Donald Hoffman's "multimodal user interface" theory: D. C. Dennett, *From Bacteria to Bach and Back: The Evolution of Minds* (New York: Norton, 2017); Amanda Gefter and Quanta Magazine, "The Case Against Reality," *Atlantic,* April 25, 2016; Donald D. Hoffman, "Sensory Experiences as Cryptic Symbols of a Multimodal User Interface," *Activitas Nervosa Superior* 52, no. 3 (September 1, 2010): 95–104; Immanuel Kant, *Critique of Pure Reason*, ed. Marcus Weigelt, trans. Friedrich Max Müller, Penguin Classics (London: Penguin Books, 2007); Plato, *Republic.*

5. Dmitriy Aronov, Rhino Nevers, and David W. Tank, "Mapping of a Non-Spatial Dimension by the Hippocampal–Entorhinal Circuit," *Nature* 543, no. 7647 (March 29, 2017): 719–22.

6. One theory is that the visual component of REM dream states may be necessary due to cortical border disputes between the senses: David Eagleman, *Livewired: The Inside Story of the Ever-Changing Brain*, ill. ed. (New York: Pantheon, 2020).

7. Examples are from: Jonathan Curot et al., "Memory Scrutinized Through Electrical Brain Stimulation: A Review of 80 Years of Experiential Phenomena," *Neuroscience and Biobehavioral Reviews* 78 (July 2017): 161–77; Kieran C. R. Fox, Lin Shi, Sori Baek, Omri Raccah, Brett L. Foster, Srijani Saha, Daniel S. Margulies, Aaron Kucyi, and Josef Parvizi, "Intrinsic Network Architecture Predicts the Effects Elicited by Intracranial Electrical Stimulation of the Human Brain," *Nature Human Behaviour* 4, no. 10 (October 2020): 1039–52; Christof Koch, "Hot or Not," *Nature Human Behaviour* 4, no. 10 (October 2020): 991–92; Aslihan Selimbeyoglu and Josef Parvizi, "Electrical Stimulation of the Human Brain:

Perceptual and Behavioral Phenomena Reported in the Old and New Literature," *Frontiers in Human Neuroscience* 4 (2010): 46.

Chapter 7: The Median Price of a Thrift-Store Bin of Evolutionary Hacks Russian-Dolled into a Watery, Salty Piñata We Call a Head

1. In August 2017, I interviewed Sydney Brenner for three days at his suite in the Shangri-La Hotel in Singapore. Brenner was one of the first people to see the proposed double-helix model of DNA, at Cambridge University, and was a longtime friend and colleague of Francis Crick, as well as a fellow Nobel laureate. He half-jokingly responded, when I asked about the evolution of brains and minds, that he had always wanted to start a scientific journal called . . . *And That's the Way the Cookie Crumbled*, with all scientists—especially, or perhaps exclusively, the biologists—required to end their manuscripts with those very words.

2. This anecdote, and its relevance to the evolution of brains and genomes, is Sydney Brenner's, told to me in person. For more on Brenner's research and life, see: Sydney Brenner, Lewis Wolpert, and Errol C. Friedberg, eds., *Sydney Brenner: A Life in Science*, rev. ed. (London: BioMed Central, 2002).

3. Michael Hopkin, "How to Make a Zombie Cockroach," *Nature* (November 29, 2007), https://doi.org/10.1038 /news.2007.312.

4. The gene, tyrosine hydroxylase, is found in many species, from us—where it is necessary for the production of dopamine, a neurotransmitter—all the way down to tiny phytoplankton. The gene can be searched for using the NIH's online Basic Local Alignment Search Tool (BLAST), which queries across a database of known genomes.

5. Craig W. Stevens, "The Evolution of Vertebrate Opioid Receptors," *Frontiers in Bioscience: A Journal and Virtual Library* 14 (January 1, 2009): 1247–69.

6. This phrase is the neuroscientist Stephen J. Smith's. For more of Smith's work, especially on the many chemical and functional similarities across cells and species, see: Ann H. Cornell-Bell, Steven M. Finkbeiner, Mark S. Cooper, and Stephen J. Smith, "Glutamate Induces Calcium Waves in Cultured Astrocytes:

Long-Range Glial Signaling," *Science* 247, no. 4941 (January 26, 1990): 470–73; Stephen J. Smith, Michael Hawrylycz, Jean Rossier, and Uygar Sümbül, "New Light on Cortical Neuropeptides and Synaptic Network Plasticity," *Current Opinion in Neurobiology*, 63 (August 1, 2020): 176–88.

7. For more on the techniques and conclusions of understanding modern genomes and behaviors from inferences about our evolutionary past, see: Paul Cisek, "Resynthesizing Behavior Through Phylogenetic Refinement," *Attention, Perception, and Psychophysics* 81, no. 7 (October 1, 2019): 2265–87; Robert M. Sapolsky, *Behave: The Biology of Humans at Our Best and Worst* (New York: Penguin Press, 2017).

8. Adam Lee et al., "Identification of an Ancient Endogenous Retrovirus, Predating the Divergence of the Placental Mammals," *Philosophical Transactions of the Royal Society B: Biological Sciences* 368, no. 1626 (September 19, 2013): 20120503.

9. Borros Arneth, "Leftovers of Viruses in Human Physiology," *Brain Structure and Function* 226, no. 6 (July 2021): 1649–58; Jens Durruthy-Durruthy et al., "The Primate-Specific Noncoding RNA HPAT5 Regulates Pluripotency during Human Preimplantation Development and Nuclear Reprogramming," *Nature Genetics* 48, no. 1 (January 2016): 44–52.

10. Sara Reardon, "Cells Hack Virus-like Protein to Communicate," *Nature* (January 11, 2018), https://doi.org/10.1038/d41586-018-00492-w; Elissa D. Pastuzyn et al., "The Neuronal Gene Arc Encodes a Repurposed Retrotransposon Gag Protein That Mediates Intercellular RNA Transfer," *Cell* 172, no. 1 (January 11, 2018): 275–288.e18.

11. The basic idea of 2R was first proposed in Susumu Ohno, *Evolution by Gene Duplication* (London: Allen and Unwin; Springer Verlag, 1970).

12. The similarity between cortical columns and their multiplicity is the basis for the computer scientist Jeff Hawkins's "thousand brains" theory: Jeff Hawkins, *A Thousand Brains: A New Theory of Intelligence* (New York: Basic Books, 2021).

13. A possible interpretation from an extremely elegant psychology experiment: see Witthoft and Winawer.

Chapter 8: Sunlight Raining Down on Gridworld

1. Arguments about thermostats and what we might now call "smart" thermostats are common in philosophy of mind, in part because thermostats are so obviously rote and mechanical but they respond behaviorally to an environment by sensing it, one of the key features of life: David John Chalmers, *The Conscious Mind: In Search of a Fundamental Theory*, Philosophy of Mind Series (New York: Oxford University Press, 1997); D. C. Dennett, *Kinds of Minds: Toward an Understanding of Consciousness* (New York: Basic Books, 1998).

2. The degree to which prediction is a consequence of life's struggles or is, in fact, the precipitating cause of life's origins is a topic of extensive debate: Andy Clark, *Surfing Uncertainty: Prediction, Action, and the Embodied Mind*, ill. ed. (Oxford: Oxford University Press, 2015); Karl Friston, "Prediction, Perception and Agency," *International Journal of Psychophysiology* 83, no. 2 (February 2012): 248–52; Jakob Hohwy, *The Predictive Mind* (Oxford: Oxford University Press, 2013).

3. Ian Sample, "'Eureka Machine' Puts Scientists in the Shade by Working Out Laws of Nature," *Guardian*, April 3, 2009, sec. Science.

4. This concept is from the physicist Erwin Schrödinger, of Schrödinger's cat fame. For more on entropy, and Schrödinger's applications of the concept to the necessary burdens of thermodynamic life: Peter Atkins, *The Laws of Thermodynamics: A Very Short Introduction* (Oxford: Oxford University Press, 2010); Eric Johnson, *Anxiety and the Equation: Understanding Boltzmann's Entropy* (Cambridge: MIT Press, 2018); Erwin Schrödinger, *"What Is Life? The Physical Aspect of the Living Cell" with "Mind and Matter" and "Autobiographical Sketches,"* Canto Classics (Cambridge: Cambridge University Press, 1992).

5. One theory is that cognition evolved because of the statistical features of the rewards as they appear in our environment: Mahi Luthra, Eduardo J. Izquierdo, and Peter M. Todd, "Cognition Evolves with the Emergence of Environmental Patchiness," in *The 2020 Conference on Artificial Life*, 450–58, online, MIT Press, 2020.

6. There have been many attempts to find an equation that perfectly describes a time of death from temperatures alone, but the variables describing the full process are myriad and confusing. A decrease of 1.5 degrees Fahrenheit per hour is a simplified heuristic used in forensics: Amy E. Maile et al., "Toward a Universal Equation to Estimate Postmortem Interval," *Forensic Science International* 272 (March 2017): 150–53; L. D. Nokes, T. Flint, J. H. Williams, and B. H. Knight, "The Application of Eight Reported Temperature-Based Algorithms to Calculate the Postmortem Interval," *Forensic Science International* 54, no. 2 (May 1992): 109–25.
7. And a former neuroscientist to boot, Dr. Eric Haseltine.
8. M. Ito, *The Cerebellum: Brain for an Implicit Self* (Upper Saddle River, New Jersey: FT Press, 2012).
9. Why is the human body the temperature that it is? Why not hotter? Colder? One theory is that mammals evolved to run in temperatures above the range that many common species of fungus can easily grow: Arturo Casadevall, "Fungi and the Rise of Mammals," *PLoS Pathogens* 8, no. 8 (August 16, 2012): e1002808.

Chapter 9: An Ante Meridiem Radio Drama

1. For a rigorous, comprehensive look at the evolution and biophysics of the eye, see: Michael F. Land and Dan-Eric Nilsson, *Animal Eyes,* Oxford Animal Biology Series (Oxford: Oxford University Press, 2012).
2. A phrase I find poetic, which I believe originated in translation from the German physiologist Ewald Hering, although I have also encountered the phrase elsewhere, including in Nick Lane's *Life Ascending*. Either way, I have borrowed it: E. Hering, *Spatial Sense and Movements of the Eye* (Baltimore, Maryland: American Academy of Optometry, 1942); Nick Lane, *Life Ascending: The Ten Great Inventions of Evolution* (New York: Norton, 2010).
3. The experimental setup where each eye is fed a different image is called "binocular rivalry": Randolph Blake and Nikos K. Logothetis, "Visual Competition," *Nature Reviews Neuroscience* 3, no. 1 (January 2002): 13–21; David Alais and Randolph Blake, eds., *Binocular Rivalry* (Cambridge, Mass: MIT Press, 2005).

4. Any good theory of consciousness should be able to explain all observed phenomena, including self-reporting of conscious perceptions, and especially the strange ones. To that end, people have reported myriad strange, blurred, or altogether confusing sights over the centuries with visual rivalry—a bit like when, on the radio, the static or faint sounds of another radio station come through, but just barely: Fatemeh Bakouie, Morteza Pishnamazi, Roxana Zeraati, and Shahriar Gharibzadeh, "Scale-Freeness of Dominant and Piecemeal Perceptions During Binocular Rivalry," *Cognitive Neurodynamics* 11, no. 4 (August 2017): 319–26; Randolph Blake, "A Primer on Binocular Rivalry, Including Current Controversies," *Brain and Mind* 2, no. 1 (April 1, 2001): 5–38; Frank Tong, Ming Meng, and Randolph Blake, "Neural Bases of Binocular Rivalry," *Trends in Cognitive Sciences* 10, no. 11 (November 2006): 502–11.

5. The majority of this chapter's arguments on visual awareness and consciousness are taken from the global neuronal workspace (GNW) theory of consciousness, by the neuroscientist Stan Dehaene, which was itself inspired by Bernard Baars's global workspace theory. GNW posits that, for example, visual experiences rise into consciousness from the journey, not the destination, as sensory information is transduced and sent across the brain from the back (occipital lobes) to the front (frontal lobes) and back again: Bernard J. Baars, *A Cognitive Theory of Consciousness* (Cambridge: Cambridge University Press, 1995) and *In the Theater of Consciousness: The Workspace of the Mind* (New York: Oxford University Press, 2001); Stanislas Dehaene, *Consciousness and the Brain: Deciphering How the Brain Codes Our Thoughts* (New York: Penguin Books, 2014).

6. One premise, or perhaps conclusion, of the GNW theory (see above) is that consciousness is, in a way, broadcast to parts of the brain.

7. A phenomenon broadly called perceptual masking, where a stimulus can inhibit perception of another; here, because an image is inhibiting perception of a previous image, it is known as "backward" masking (as opposed to "simultaneous" or "forward" masking). Of note, the ability of a stimulus to retroactively mask an earlier one is strong evidence that

consciousness, or at least visual awareness, is not just an on-air stream of incoming sensory inputs but, instead, a kind of puzzled-together set of sensory guesses with physical, or logical, rules about priority—just like a broadcast, which, as GNW argues, requires that the message be sent and received: Bruno G. Breitmeyer and Haluk Öğmen, *Visual Masking: Time Slices Through Conscious and Unconscious Vision*, Oxford Psychology Series 41 (Oxford: Oxford University Press, 2006); Stanislas Dehaene and Jean-Pierre Changeux, "Experimental and Theoretical Approaches to Conscious Processing," *Neuron* 70, no. 2 (April 28, 2011): 200–227; Johannes J. Fahrenfort, Jonathan van Leeuwen, Christian N. L. Olivers, and Hinze Hogendoorn, "Perceptual Integration without Conscious Access," *Proceedings of the National Academy of Sciences* 114, no. 14 (April 4, 2017): 3744–49; Antonino Raffone, Narayanan Srinivasan, and Cees van Leeuwen, "Perceptual Awareness and Its Neural Basis: Bridging Experimental and Theoretical Paradigms," *Philosophical Transactions of the Royal Society B: Biological Sciences* 369, no. 1641 (May 5, 2014): 20130203.

Chapter 10: A Small Town with Too Much Food

1. This chapter draws heavily from multiple books on evolution, chemistry, neuroscience, and theories on the origins of life in order to tell a tale about the single, unbroken line connecting the earliest life on Earth to the moments of Anna's surgery. I took a few stylistic, creative liberties: as in evolution by natural selection, which can never restart, the chapter has no line breaks; and many of the concepts, once introduced, build off of instead of replace one another. This leads to awkward, inelegant phrases like "superlong rewrapped net-oar huts" and "spring-loaded wind chime traps" with the analog that some pieces of us, too, can be equally if not more awkward in composition (the mammalian eye, for example, which appears to be built inside out with all the light-sensitive cells in the back). For the sake of coherence, a simple story is presented throughout, as if the events of its narrative happened one at a time or in an ordained, linear order. Of course, the real story is fantastically more complex and there is rarely a scientific consensus about the details, many of which were lost sometime in the last few billion years, give or take: A. Graham

Cairns-Smith, *Evolving the Mind: On the Nature of Matter and the Origin of Consciousness* (Cambridge: Cambridge University Press, 1998); Todd E. Feinberg and Jon M. Mallatt, *The Ancient Origins of Consciousness: How the Brain Created Experience* (Cambridge: MIT Press, 2016); Peter Godfrey-Smith, *Metazoa: Animal Life and the Birth of the Mind* (New York: Farrar, Straus and Giroux, 2020); Eric R. Kandel, James H. Schwartz, Thomas M. Jessell, Steven A. Siegelbaum, and A. J. Hudspeth, eds., *Principles of Neural Science*, 5th ed. (New York: McGraw-Hill, 2012); Nick Lane, *Life Ascending: The Ten Great Inventions of Evolution*, ill. ed. (New York: Norton, 2010); *Oxygen: The Molecule That Made the World*, rev. impression, Oxford Landmark Science Series (Oxford: Oxford University Press, 2016); *The Vital Question: Energy, Evolution, and the Origins of Complex Life* (New York: Norton, 2016); *Power, Sex, Suicide: Mitochondria and the Meaning of Life*, 2nd ed., Oxford Landmark Science Series (Oxford: Oxford University Press, 2018).

Chapter 11: The Arbiter of Elegance

1. Gilbert Bagnani, *Arbiter of Elegance: A Study of the Life and Works of C. Petronius* (University of Toronto Press, 1958).

2. Petronius, *The Satyricon* (New York: Meridian, 1994).

3. For example, a rare Japanese flower's genome is around fifty times larger than the human genome: Jaume Pellicer et al., "The Largest Eukaryotic Genome of Them All?" *Botanical Journal of the Linnean Society* 164, no. 1 (September 1, 2010): 10–15.

4. The neuroscientist Rodolfo Llinás argues that aspects of the sea squirt life cycle—specifically, that its nervous system dissolves around the same time as it becomes stationary—are evidence that "[t]he generation of movement and the generation of mindness are deeply related; they are in fact different thoughts of the same process": Rodolfo R. Llinás, *I of the Vortex*.

5. As Llinás (*I of the Vortex*) has starkly put it, "you either move or drool."

6. Because of the massive metabolism required for flight and echolocation, not to mention the base mammalian extravagances, bats have reduced immune-system sensitivity, so as to not attack their own cells; likely, this compromise is one of many reasons

bats are veritable sponges of virus, carrying the rabies, Ebola, corona, and Marburg viruses, to name a few: Arinjay Banerjee, Michelle L. Baker, Kirsten Kulcsar, Vikram Misra, Raina Plowright, and Karen Mossman, "Novel Insights into Immune Systems of Bats," *Frontiers in Immunology* 11 (2020): 26.

7. Rich Pang, Benjamin J. Lansdell, and Adrienne L. Fairhall, "Dimensionality Reduction in Neuroscience," *Current Biology* 26, no. 14 (July 25, 2016): R656–60.

8. Again, this idea aligns with arguments made by Llinás (*I of the Vortex*): "In my view, from its evolutionary inception, mindness is internalisation of movement."

9. Patrick House, "What Is Elegance in Science?," *New Yorker*, August 17, 2015.

Chapter 12: Swinging Through Ancient Trees While Standing Still and Hearing Voices

1. This is a theory, only one among many, about the evolution of human language: Jacques Prieur, Stéphanie Barbu, Catherine Blois-Heulin, and Alban Lemasson, "The Origins of Gestures and Language: History, Current Advances and Proposed Theories," *Biological Reviews* 95, no. 3 (2020): 531–54.

2. Of course, more vocal-learning species might be discovered one day. For a current review, see: Erich D. Jarvis, "Evolution of Vocal Learning and Spoken Language," *Science* 366, no. 6461 (October 4, 2019): 50–54.

3. Jonathan D. Breshears, Derek G. Southwell, and Edward F. Chang, "Inhibition of Manual Movements at Speech Arrest Sites in the Posterior Inferior Frontal Lobe," *Neurosurgery* 85, no. 3 (September 1, 2019): E496–501.

4. It is interesting that some of Oliver Sacks's Parkinsonian, catatonic patients, when given a drug that temporarily relieved their immobility, reported an inability to *both move and think* while catatonic: Oliver Sacks, *Awakenings* (New York: Vintage Books, 1999).

5. Daniel B. Smith, *Muses, Madmen, and Prophets: Hearing Voices and the Borders of Sanity* (New York: Penguin, 2008).

6. Anne-Kathrin J. Fett, Imke L. J. Lemmers-Jansen, and Lydia Krabbendam, "Psychosis and Urbanicity: A Review of the Recent Literature from Epidemiology to Neurourbanism," *Current Opinion in Psychiatry* 32, no. 3 (May 2019): 232–41.

7. Eduard Einstein, Lucia Joyce, John Russell.

8. Shannon Wiltsey Stirman and James W. Pennebaker, "Word Use in the Poetry of Suicidal and Nonsuicidal Poets," *Psychosomatic Medicine* 63, no. 4 (July 2001): 517–22.

9. Charity J. Morgan et al., "Thought Disorder in Schizophrenia and Bipolar Disorder Probands, Their Relatives, and Nonpsychiatric Controls," *Schizophrenia Bulletin* 43, no. 3 (May 2017): 523–35.

10. Thomas S. Kuhn and Ian Hacking, *The Structure of Scientific Revolutions*, 4th ed. (Chicago: University of Chicago Press, 2012).

Chapter 13: Endeavoring to Grow Wings

1. Frigyes Karinthy and Oliver Sacks, *A Journey Round My Skull*, trans. Vernon Duckworth Barker (New York: NYRB Classics, 2008).

2. Patrik Vuilleumier, "Anosognosia: The Neurology of Beliefs and Uncertainties," *Cortex* 40, no. 1 (January 1, 2004): 9–17.

3. Here Koch is referring to mental time travel, which is the ability to reconstruct personal events from memory as well as imagine possible scenarios into the future—in other words, our ability to both remember and daydream. The centrality of this ability to the workings of consciousness and the evolution of the human mind is an ongoing debate: Donna Rose Addis, "Mental Time Travel? A Neurocognitive Model of Event Simulation," *Review of Philosophy and Psychology* 11, no. 2 (June 1, 2020): 233–59; Thomas Suddendorf, Donna Rose Addis, and Michael C. Corballis, "Mental Time Travel and the Shaping of the Human Mind," *Philosophical Transactions of the Royal Society B: Biological Sciences* 364, no. 1521 (May 12, 2009): 1317–24; Thomas Suddendorf and Michael C. Corballis, "The Evolution of Foresight: What Is Mental Time Travel, and Is It Unique to Humans?" *Behavioral and Brain Sciences* 30, no. 3 (June 2007): 299–313.

Chapter 14: The North African Rhino of Charismatic Megaquale

1. "Quale" is a philosophical term, often defined as the subjective or qualitative properties of experiences or conscious states. The plural form is "qualia," from the Latin for "of what kind."

2. From the novelist Dorothy Richardson: Paul Tiessen, "A Comparative Approach to the Form and Function of Novel and Film: Dorothy Richardson's Theory of Art," *Literature/Film Quarterly* 3, no. 1 (1975): 83–90.

3. Caveat: I am no geologist or geophysicist. I pieced these possible events together from cursory readings on geology, magnetic fields, and planetary motion.

4. A classic work in neuroscience: Michael S. Gazzaniga, *The Consciousness Instinct: Unraveling the Mystery of How the Brain Makes the Mind* (New York: Farrar, Straus and Giroux, 2018). See also Christof Koch and Patrick House, "Brain Bridging," *Nature*, August 26, 2020, d41586-020-02469-0; David Wolman, "The Split Brain: A Tale of Two Halves," *Nature* 483, no. 7389 (March 1, 2012): 260–63.

5. The arguments in this chapter regarding phi, or Φ, are pulled from postulates of "integrated information theory" (IIT). Early aspects of IIT were developed by the neuroscientist Guilio Tononi and elaborated on over decades of collaborations with Christof Koch and others. The descriptions in this chapter are pulled from extensive personal conversations with Koch and Tononi, as well as from: Giulio Tononi and Christof Koch, "Consciousness: Here, There and Everywhere?" *Philosophical Transactions of the Royal Society B: Biological Sciences* 370, no. 1668 (May 19, 2015): 20140167; Christof Koch, *The Feeling of Life Itself: Why Consciousness Is Widespread but Can't Be Computed,* ill. ed. (Cambridge: MIT Press, 2019) and *Consciousness: Confessions of a Romantic Reductionist* (Cambridge: MIT Press, 2012); Marcello Massimini and Giulio Tononi, *Sizing Up Consciousness: Towards an Objective Measure of the Capacity for Experience* (Oxford: Oxford University Press, 2018).

6. This part of the chapter, which could easily be called the "Malarone and Monks" section, is a stream-of-consciousness homage to chapter 34 of Julio Cortázar's "Hopscotch," which alternates each sentence between two minds, to the delight and confusion of any first-time reader. The idea that there may one day be a physics of thought and that its topography could be learned, perhaps, to predict where thoughts may go is drawn from personal conversation with psychologist Diana Tamir, and from: Judith N. Mildner and Diana I.

Tamir, "Spontaneous Thought as an Unconstrained Memory Process," *Trends in Neurosciences* 42, no. 11 (November 2019): 763–77.

Chapter 15: An Itsy-Bitsy Teeny-Weeny Quantum-Dot-Like Non-Machiney

1. Referring, probably, to the Jacquard loom: F.W.H. Myers, "Multiplex personality," *Proceedings of the Society for Psychical Research* 4 (1887): 496–514.
2. This chapter, unlike many others, is based almost exclusively on a single theory of consciousness: the orchestrated objective reduction (Orch OR) theory, sometimes known as the Penrose-Hameroff theory. Roger Penrose, a physicist, and Stuart Hameroff, an anesthesiologist, outlined these ideas in a series of publications in the last few decades. The theory beautifully combines discoveries from mathematics, including Gödel's incompleteness theorems, with arguments from the philosophy of computation, chemistry, and cellular biology. For an overview, see: Stuart Hameroff and Roger Penrose, "Consciousness in the Universe: A Review of the 'Orch OR' Theory," *Physics of Life Reviews* 11, no. 1 (March 1, 2014): 39–78; Roger Penrose, *Shadows of the Mind: A Search for the Missing Science of Consciousness*, repr. ed. (Oxford: Oxford University Press, 1996).
3. For scale, this means that, with an enlarging ray, if you expanded both a human hair and a microtubule, such that the human hair grew to one mile in diameter, the microtubule would be only about twenty inches across.
4. For debates on and criticism of Orch OR, see: Stuart Hameroff and Roger Penrose, "Reply to Seven Commentaries on 'Consciousness in the Universe: Review of the "Orch OR" Theory,'" *Physics of Life Reviews* 11, no. 1 (March 1, 2014): 94–100; Christof Koch and Klaus Hepp, "Quantum Mechanics in the Brain," *Nature* 440, no. 7084 (March 2006): 61.

Chapter 16: A Make-Believe Parasite with No Legs and Places to Go

1. Ryan P. Dalton and Tom Roseberry, "Studying the Superhuman," *Scientific American* (September 4, 2019).

2. The idea here is that one day we may be able to combine consciousness by connecting brains sufficiently well, much as our two hemispheres are bridged at birth by a natural connection between them: Michael S. Gazzaniga, *Tales from Both Sides of the Brain: A Life in Neuroscience*, ill. ed. (New York: Ecco, 2015); Koch and House, "Brain Bridging."
3. Witthoft and Winawer.
4. I warned you. I'm a parasite guy.
5. This idea is related to panpsychism, the theory that all things have consciousness and that our minds simply corral a lot of these fundamental particles of consciousness, like herding dogs corral sheep. The theory is popular, despite being at odds with common sense, and is at the center of intense debates, though they often take the form of debates about the structure of debate: Keith Frankish and Aeon, "Why Panpsychism Is Probably Wrong," *Atlantic,* September 20, 2016; Philip Goff, *Galileo's Error: Foundations for a New Science of Consciousness* (New York: Pantheon Books, 2019) and "Panpsychism Is Crazy, but It's Also Most Probably True," Aeon, March 1, 2017.
6. Usually, a raindrop forms around a bacteria trapped high in the sky, or maybe around a speck of dust.

Chapter 17: A Sex-Starved Cricket Sculpting in Time

1. Broadly, this kind of message within a nervous system is known as, depending on the details, an "efference copy," "corollary discharge," or a "reafferent suppression": Matasaburo Fukutomi and Bruce A. Carlson, "A History of Corollary Discharge: Contributions of Mormyrid Weakly Electric Fish," *Frontiers in Integrative Neuroscience* 14 (2020): 42; Hans Straka, John Simmers, and Boris P. Chagnaud, "A New Perspective on Predictive Motor Signaling," *Current Biology* 28, no. 5 (March 2018): R232–43.
2. The chirp is a male cricket's come-hither: R. K. Murphey and John Palka, "Efferent Control of Cricket Giant Fibres," *Nature* 248, no. 5445 (March 1974): 249–51; Harald Nocke, "Physiological Aspects of Sound Communication in Crickets (Gryllus Campestris L.)," *Journal of Comparative Physiology* 80, no. 2 (June 1, 1972): 141–62.
3. For general discussions of eye movements and what can be

learned from them: Mary Hayhoe and Dana Ballard, "Eye Movements in Natural Behavior," *Trends in Cognitive Sciences* 9, no. 4 (April 2005): 188–94; Grace Lindsay et al., "Episode 25: What Can Eye Movements Tell Us About the Mind?," October 1, 2017, in *Unsupervised Thinking*, podcast.

4. Yuki Kishita, Hiroshi Ueda, and Makio Kashino, "Eye and Head Movements of Elite Baseball Players in Real Batting," *Frontiers in Sports and Active Living* (January 29, 2020); M. Land, N. Mennie, and J. Rusted, "The Roles of Vision and Eye Movements in the Control of Activities of Daily Living," *Perception* 28, no. 11 (1999): 1311–28; Matthew D. Shank and Kathleen M. Haywood, "Eye Movements While Viewing a Baseball Pitch," *Perceptual and Motor Skills* 64, no. 3 suppl. (June 1, 1987): 1191–97.

5. There is an analog for this in the early days of the television industry, where compressed television signals transmitted only the difference in pixels from one frame to the next. For example, if the camera remained still, the background objects—say, paintings on the walls of the bar in *Cheers*—would not be transmitted wholly each frame; only the changes would be sent from one frame to the next. Likewise, certain upper echelons of the brain are only receiving messages about what has changed relative to previous sensory input.

Chapter 18: A Small (or Large) Learning Machine Made Out of Words

1. This argument, and many others throughout this chapter, are drawn from personal conversation with the psychologist Diana Tamir and from Mildner and Tamir.

2. Thomas Cleary, *Code of the Samurai: A Modern Translation of the Bushido Shoshinshu of Taira Shigesuke* (Hong Kong: Tuttle Publishing, 1999).

3. From Kipling's poem "Boots": "'Tain't—so—bad—by—day because o' company, / But night—brings—long—strings—o' forty thousand million / Boots—boots—boots—boots—movin' up an' down again. / There's no discharge in the war!"

4. If we started using mind-reading machines in court, could we really tell the difference between intention and fantasy? How would society change if we had full access to others' thoughts? Is someone "bad" by virtue of the composition of their thoughts,

or might the definition of "good" come to be the suppression of or ability to avoid acting on the evil, worst thoughts?

5. This estimate is from the psychologist David Gilbert.

6. To be fair, we don't trust humans to drive either until they have had at least sixteen years of training in the statistics of how the world works.

7. "Off Road, but Not Offline: How Simulation Helps Advance Our Waymo Driver," Waymo, April 28, 2020.

8. "OpenAI Five."

9. Jane Wang, Zeb Kurth-Nelson, and Matt Botvinick, "Prefrontal Cortex as a Meta-reinforcement Learning System," *Deepmind*, May 14, 2018.

10. H. Freyja Ólafsdóttir, Daniel Bush, and Caswell Barry, "The Role of Hippocampal Replay in Memory and Planning," *Current Biology* 28, no. 1 (January 8, 2018): R37–50.

11. Robert Stickgold, April Malia, Denise Maguire, David Roddenberry, and Margaret O'Connor, "Replaying the Game: Hypnagogic Images in Normals and Amnesics," *Science* 290, no. 5490 (October 13, 2000): 350–53.

12. An effect known as "game transfer phenomena": Angelica B. Ortiz de Gortari and Mark D. Griffiths, "Game Transfer Phenomena and Its Associated Factors: An Exploratory Empirical Online Survey Study," *Computers in Human Behavior* 51 (October 1, 2015): 195–202.

13. This beautiful idea is the computer scientist Blaise Agüera y Arcas's, told to me in personal conversation.

14. This is Christof Koch's argument, from: Koch, *The Feeling of Life Itself*.

Chapter 19: Not That Hard

1. Toward the end of the twentieth century, the philosopher David Chalmers introduced the concept of the "easy" and "hard" problems of consciousness, a debate that still rages— on terms, on proposed solutions, on edge cases: Ned Block, "Comparing the Major Theories of Consciousness," in *The Cognitive Neurosciences*, 4th ed. (Cambridge: MIT Press, 2009): 1111–22; David Chalmers, "Facing Up to the Problem of Consciousness," and *The Character of Consciousness*, ill. ed. (New York: Oxford University Press, 2010); Daniel C. Dennett, *Consciousness Explained* (Boston: Back Bay

Books, 1992) and "Facing Up to the Hard Question of Consciousness," *Philosophical Transactions of the Royal Society B: Biological Sciences* 373, no. 1755 (September 19, 2018): 20170342.

2. Anne-Laure Lemaitre, Marion Luyat, and Gilles Lafargue, "Individuals with Pronounced Schizotypal Traits Are Particularly Successful in Tickling Themselves," *Consciousness and Cognition* 41 (April 1, 2016): 64–71.

3. Susan Dominus, "Could Conjoined Twins Share a Mind?" *New York Times*, May 25, 2011, sec. Magazine.

4. This chapter's proposed answer to the *H. caecus* thought experiment is a combination of ideas from a conversation between me and Alexander Wissner-Gross, a physicist and computer scientist. For more of Wissner-Gross's work on inferences made about hard-to-detect things, see: Alexander Wissner-Gross, "A New Equation for Intelligence," TED Talk, 2014.

Bibliography

Addis, Donna Rose. "Mental Time Travel? A Neurocognitive Model of Event Simulation." *Review of Philosophy and Psychology* 11, no. 2 (June 1, 2020): 233–59. https://doi.org/10.1007/s13164 -020-00470-0.

Alais, David, and Randolph Blake, eds. *Binocular Rivalry*. Cambridge: MIT Press, 2005.

Arneth, Borros. "Leftovers of Viruses in Human Physiology." *Brain Structure and Function* 226, no. 6 (July 2021): 1649–58. https:// doi.org/10.1007/s00429-021-02306-8.

Aronov, Dmitriy, Rhino Nevers, and David W. Tank. "Mapping of a Non-Spatial Dimension by the Hippocampal–Entorhinal Circuit." *Nature* 543, no. 7647 (March 29, 2017): 719–22. https:// doi.org/10.1038/nature21692.

Atkins, Peter. *The Laws of Thermodynamics: A Very Short Introduction*. Oxford: Oxford University Press, 2010.

Baars, Bernard J. *A Cognitive Theory of Consciousness*. Repr. ed. Cambridge: Cambridge University Press, 1995.

———. *In the Theater of Consciousness: The Workspace of the Mind*. Oxford: Oxford University Press, 2001.

"The Bagatelle Wizard Instead of the Pinball Wizard." National Museum of American History, October 31, 2012. https:// americanhistory.si.edu/blog/2012/10/the-bagatelle-wizard-instead -of-the-pinball-wizard.html.

Bagnani, Gilbert. *Arbiter of Elegance: A Study of the Life and Works of C. Petronius*. Toronto: University of Toronto Press, 1958.

Baird, Benjamin, Stephen LaBerge, and Giulio Tononi. "Two-Way Communication in Lucid REM Sleep Dreaming." *Trends in Cognitive Sciences* 25, no. 6 (June 1, 2021): 427–28. https://doi .org/10.1016/j.tics.2021.04.004.

Bakouie, Fatemeh, Morteza Pishnamazi, Roxana Zeraati, and Shahriar Gharibzadeh. "Scale-Freeness of Dominant and Piece-meal Perceptions during Binocular Rivalry." *Cognitive Neurody-*

namics 11, no. 4 (August 2017): 319–26. https://doi.org/10.1007 /s11571-017-9434-4.

Balzac, Honoré de. *Treatise on Elegant Living*. Translated by Napoleon Jeffries. 2nd ed. Cambridge, Mass.: Wakefield Press, 2010.

Banerjee, Arinjay, Michelle L. Baker, Kirsten Kulcsar, Vikram Misra, Raina Plowright, and Karen Mossman. "Novel Insights into Immune Systems of Bats." *Frontiers in Immunology* 11 (2020): 26. https://doi.org/10.3389/fimmu.2020.00026.

Bernstein, N. *The Coordination and Regulation of Movements*. Oxford: Pergamon Press, 1967.

Bernstein, Nicholai A. *Dexterity and Its Development*. Edited by Mark L. Latash and Michael T. Turvey. New York: Psychology Press, 2015.

Blackiston, Douglas J., Elena Silva Casey, and Martha R. Weiss. "Retention of Memory through Metamorphosis: Can a Moth Remember What It Learned as a Caterpillar?" *PloS One* 3, no. 3 (March 5, 2008): e1736. https://doi.org/10.1371/journal.pone .0001736.

Blackmore, Susan. *Consciousness: A Very Short Introduction*. New York: Oxford University Press, 2017.

Blake, Randolph. "A Primer on Binocular Rivalry, Including Current Controversies." *Brain and Mind* 2, no. 1 (April 1, 2001): 5–38. https://doi.org/10.1023/A:1017925416289.

Blake, Randolph, and Nikos K. Logothetis. "Visual Competition." *Nature Reviews Neuroscience* 3, no. 1 (January 2002): 13–21. https://doi.org/10.1038/nrn701.

Block, Ned. "Comparing the Major Theories of Consciousness." In *The Cognitive Neurosciences*, 4th ed., edited by Michael S. Gazzaniga, 1111–22. Cambridge: MIT Press, 2009.

Bray, Dennis. *Wetware: A Computer in Every Living Cell*. Ill. ed. New Haven: Yale University Press, 2011.

Breitmeyer, Bruno G., and Haluk Öğmen. *Visual Masking: Time Slices Through Conscious and Unconscious Vision*. 2nd ed. Oxford Psychology Series 41. Oxford: Oxford University Press, 2006.

Brenner, Sydney, Lewis Wolpert, and Errol C. Friedberg, eds. *Sydney Brenner: A Life in Science*. Rev. ed. London: BioMed Central, 2002.

Breshears, Jonathan D., Derek G. Southwell, and Edward F. Chang. "Inhibition of Manual Movements at Speech Arrest Sites in the Posterior Inferior Frontal Lobe." *Neurosurgery* 85, no. 3 (September 1, 2019): E496–501. https://doi.org/10.1093/neuros/nyy592.

Buonomano, Dean. *Your Brain Is a Time Machine: The Neuroscience and Physics of Time*. New York: Norton, 2017.

Buzsáki, György. *The Brain from Inside Out*. Repr. ed. Oxford: Oxford University Press, 2021.

———. *Rhythms of the Brain*. Oxford: Oxford University Press, 2006.

Cairns-Smith, A. Graham. *Evolving the Mind: On the Nature of Matter and the Origin of Consciousness*. Cambridge: Cambridge University Press, 1998.

Casadevall, Arturo. "Fungi and the Rise of Mammals." *PLoS Pathogens* 8, no. 8 (August 16, 2012): e1002808. https://doi.org/10.1371/journal.ppat.1002808.

Chalmers, David J. *The Character of Consciousness*. Ill. ed. New York: Oxford University Press, 2010.

———. *The Conscious Mind: In Search of a Fundamental Theory*. Philosophy of Mind Series. New York: Oxford University Press, 1997.

———. "Facing Up to the Problem of Consciousness." *Journal of Consciousness Studies* 2, no. 3 (1995): 200–219.

Cisek, Paul. "Resynthesizing Behavior through Phylogenetic Refinement." *Attention, Perception, and Psychophysics* 81, no. 7 (October 1, 2019): 2265–87. https://doi.org/10.3758/s13414-019-01760-1.

Clark, Andy. *Surfing Uncertainty: Prediction, Action, and the Embodied Mind*. Ill. ed. Oxford: Oxford University Press, 2015.

Cleary, Thomas. *Code of the Samurai: A Modern Translation of the Bushido Shoshinshu of Taira Shigesuke*. Hong Kong: Tuttle Publishing, 1999.

Corballis, Michael C. *The Wandering Mind: What the Brain Does When You're Not Looking*. Chicago: University of Chicago Press, 2015.

Cornell-Bell, Ann H., Steven M. Finkbeiner, Mark S. Cooper, and Stephen J. Smith. "Glutamate Induces Calcium Waves in Cultured Astrocytes: Long-Range Glial Signaling." *Science* 247, no. 4941 (January 26, 1990): 470–73. https://doi.org/10.1126/science.1967852.

Crick, Francis. *Astonishing Hypothesis: The Scientific Search for the Soul*. Repr. ed. London: Scribner, 1995.

Curot, Jonathan, et al. "Memory Scrutinized Through Electrical
　　Brain Stimulation: A Review of 80 Years of Experiential Phenom-
　　ena." *Neuroscience and Biobehavioral Reviews* 78 (July 2017):
　　161–77. https://doi.org/10.1016/j.neubiorev.2017.04.018.

Dalton, Ryan P., and Tom Roseberry. "Studying the Superhu-
　　man." *Scientific American.* September 4, 2019. https://www
　　.scientificamerican.com/article/studying-the-superhuman/.

Damasio, Antonio. *Descartes' Error: Emotion, Reason, and the Human
　　Brain.* Ill. ed. London: Penguin Books, 2005.

———. *The Feeling of What Happens: Body and Emotion in the Mak-
　　ing of Consciousness.* San Diego: Mariner Books, 2000.

Dehaene, Stanislas. *Consciousness and the Brain: Deciphering How
　　the Brain Codes Our Thoughts.* New York: Penguin Books, 2014.

Dehaene, Stanislas, and Jean-Pierre Changeux. "Experimental and
　　Theoretical Approaches to Conscious Processing." *Neuron* 70,
　　no. 2 (April 28, 2011): 200–27. https://doi.org/10.1016/j.neuron
　　.2011.03.018.

Dennett, Daniel C. *Consciousness Explained.* Boston: Back Bay
　　Books, 1992.

———. "Facing Up to the Hard Question of Consciousness." *Phil-
　　osophical Transactions of the Royal Society B: Biological Sciences*
　　373, no. 1755 (September 19, 2018): 20170342. https://doi.org
　　/10.1098/rstb.2017.0342.

———. *From Bacteria to Bach and Back: The Evolution of Minds.*
　　New York: Norton, 2017.

———. *Kinds of Minds: Toward an Understanding of Consciousness.*
　　New York: Basic Books, 1998.

Dominus, Susan. "Could Conjoined Twins Share a Mind?" *New York
　　Times,* May 25, 2011, sec. Magazine. https://www.nytimes.com
　　/2011/05/29/magazine/could-conjoined-twins-share-a-mind.html.

Duinen, Hiske van, and Simon C. Gandevia. "Constraints for
　　Control of the Human Hand." *The Journal of Physiology* 589,
　　no. Pt 23 (December 1, 2011): 5583–93. https://doi.org/10.1113
　　/jphysiol.2011.217810.

Durruthy-Durruthy et al. "The Primate-Specific Noncoding RNA
　　HPAT5 Regulates Pluripotency During Human Preimplantation
　　Development and Nuclear Reprogramming." *Nature Genetics* 48,
　　no. 1 (January 2016): 44–52. https://doi.org/10.1038/ng.3449.

Eagleman, David. *Livewired: The Inside Story of the Ever-Changing
　　Brain.* Ill. ed. New York: Pantheon, 2020.

Faber, Donald S., and Alberto E. Pereda. "Two Forms of Electrical Transmission Between Neurons." *Frontiers in Molecular Neuroscience* 11 (November 21, 2018): 427. https://doi.org/10.3389/fnmol.2018.00427.

Fahrenfort, Johannes J., Jonathan van Leeuwen, Christian N. L. Olivers, and Hinze Hogendoorn. "Perceptual Integration without Conscious Access." *Proceedings of the National Academy of Sciences* 114, no. 14 (April 4, 2017): 3744–49. https://doi.org/10.1073/pnas.1617268114.

Feinberg, Todd E., and Jon M. Mallatt. *The Ancient Origins of Consciousness: How the Brain Created Experience*. Cambridge: MIT Press, 2016.

Ferrarelli, Fabio, Marcello Massimini, Simone Sarasso, Adenauer Casali, Brady A. Riedner, Giuditta Angelini, Giulio Tononi, and Robert A. Pearce. "Breakdown in Cortical Effective Connectivity During Midazolam-Induced Loss of Consciousness." *Proceedings of the National Academy of Sciences of the United States of America* 107, no. 6 (February 9, 2010): 2681–86. https://doi.org/10.1073/pnas.0913008107.

Fett, Anne-Kathrin J., Imke L. J. Lemmers-Jansen, and Lydia Krabbendam. "Psychosis and Urbanicity: A Review of the Recent Literature from Epidemiology to Neurourbanism." *Current Opinion in Psychiatry* 32, no. 3 (May 2019): 232–41. https://doi.org/10.1097/YCO.0000000000000486.

Fox, Kieran C. R., Lin Shi, Sori Baek, Omri Raccah, Brett L. Foster, Srijani Saha, Daniel S. Margulies, Aaron Kucyi, and Josef Parvizi. "Intrinsic Network Architecture Predicts the Effects Elicited by Intracranial Electrical Stimulation of the Human Brain." *Nature Human Behaviour* 4, no. 10 (October 2020): 1039–52. https://doi.org/10.1038/s41562-020-0910-1.

Frankish, Keith, and Aeon. "Why Panpsychism Is Probably Wrong." *Atlantic*, September 20, 2016. https://www.theatlantic.com/science/archive/2016/09/panpsychism-is-wrong/500774/.

Friston, Karl. "Am I Self-Conscious? (Or Does Self-Organization Entail Self-Consciousness?)." *Frontiers in Psychology* 9 (2018): 579. https://doi.org/10.3389/fpsyg.2018.00579.

———. "The Free-Energy Principle: A Unified Brain Theory?" *Nature Reviews Neuroscience* 11, no. 2 (February 2010): 127–38. https://doi.org/10.1038/nrn2787.

———. "A Free Energy Principle for Biological Systems." *Entropy*

14, no. 11 (November 2012): 2100–21. https://doi.org/10.3390
/e14112100.

———. "Life as We Know It." *Journal of the Royal Society Interface*
10, no. 86 (September 6, 2013): 20130475. https://doi.org/10
.1098/rsif.2013.0475.

———. "Prediction, Perception and Agency." *International Journal of
Psychophysiology* 83, no. 2 (February 2012): 248–52. https://doi
.org/10.1016/j.ijpsycho.2011.11.014.

Fukutomi, Matasaburo, and Bruce A. Carlson. "A History of Corollary Discharge: Contributions of Mormyrid Weakly Electric
Fish." *Frontiers in Integrative Neuroscience* 14 (2020): 42. https://
doi.org/10.3389/fnint.2020.00042.

Fyhn, Marianne, Torkel Hafting, Alessandro Treves, May-Britt
Moser, and Edvard I. Moser. "Hippocampal Remapping and
Grid Realignment in Entorhinal Cortex." *Nature* 446, no. 7132
(March 2007): 190–94. https://doi.org/10.1038/nature05601.

Gazzaniga, Michael S. *The Consciousness Instinct: Unraveling the
Mystery of How the Brain Makes the Mind.* New York: Farrar,
Straus and Giroux, 2018.

———. *Tales from Both Sides of the Brain: A Life in Neuroscience.* Ill.
ed. New York: Ecco, 2015.

Gefter, Amanda, and Quanta Magazine. "The Case Against Reality."
Atlantic, April 25, 2016. https://www.theatlantic.com/science
/archive/2016/04/the-illusion-of-reality/479559/.

Glynn, Ian. *Elegance in Science: The Beauty of Simplicity.* Oxford:
Oxford University Press, 2013.

Godfrey-Smith, Peter. *Metazoa: Animal Life and the Birth of the
Mind.* New York: Farrar, Straus and Giroux, 2020.

Goff, Philip. *Galileo's Error: Foundations for a New Science of Consciousness.* New York: Pantheon Books, 2019.

———. "Panpsychism Is Crazy, but It's Also Most Probably True."
Aeon, March 1, 2017. https://aeon.co/ideas/panpsychism-is-crazy
-but-its-also-most-probably-true.

Graziano, Michael S. A. *Rethinking Consciousness: A Scientific Theory of Subjective Experience.* New York: Norton, 2019.

Halley, J. M. "Ecology, Evolution and 1/f-Noise." *Trends in Ecology
and Evolution* 11, no. 1 (1996): 33–37. https://doi.org/10.1016
/0169–5347(96)81067-6.

Hameroff, Stuart, and Roger Penrose. "Consciousness in the Universe: A Review of the 'Orch OR' Theory." *Physics of Life Reviews*

11, no. 1 (March 1, 2014): 39–78. https://doi.org/10.1016/j.plrev
.2013.08.002.

———. "Reply to Seven Commentaries on 'Consciousness in the
Universe: Review of the "Orch OR" Theory.'" *Physics of Life
Reviews* 11, no. 1 (March 1, 2014): 94–100. https://doi.org/10
.1016/j.plrev.2013.11.013.

Hawkins, Jeff. *A Thousand Brains: A New Theory of Intelligence*. New
York: Basic Books, 2021.

Hayhoe, Mary, and Dana Ballard. "Eye Movements in Natural
Behavior." *Trends in Cognitive Sciences* 9, no. 4 (April 2005):
188–94. https://doi.org/10.1016/j.tics.2005.02.009.

Hazen, Robert M. *The Story of Earth: The First 4.5 Billion Years, from
Stardust to Living Planet*. Repr. ed. New York: Penguin Books, 2013.

Heider, Fritz, and Marianne Simmel. "An Experimental Study of
Apparent Behavior." *American Journal of Psychology* 57, no. 2
(1944): 243–59. https://doi.org/10.2307/1416950.

Held, Richard, Yuri Ostrovsky, Beatrice de Gelder, Tapan Gandhi,
Suma Ganesh, Umang Mathur, and Pawan Sinha. "The Newly
Sighted Fail to Match Seen with Felt." *Nature Neuroscience* 14,
no. 5 (May 2011): 551–53. https://doi.org/10.1038/nn.2795.

Hoffman, Donald D. "Sensory Experiences as Cryptic Symbols of
a Multimodal User Interface." *Activitas Nervosa Superior* 52,
no. 3 (September 1, 2010): 95–104. https://doi.org/10.1007
/BF03379572.

Hohwy, Jakob. *The Predictive Mind*. Oxford: Oxford University Press,
2013.

Hopkin, Michael. "How to Make a Zombie Cockroach." *Nature*,
November 29, 2007. https://doi.org/10.1038/news.2007.312.

House, Patrick. "What Is Elegance in Science?" *New Yorker*.
Accessed October 10, 2021. https://www.newyorker.com/tech
/annals-of-technology/what-is-elegance-in-science.

———. "What People Cured of Blindness See." *New Yorker*, August
28, 2014. http://www.newyorker.com/tech/elements/people
-cured-blindness-see.

House, Patrick K., Ajai Vyas, and Robert Sapolsky. "Predator Cat
Odors Activate Sexual Arousal Pathways in Brains of Toxoplasma
Gondii Infected Rats." *PLoS One* 6, no. 8 (August 17, 2011):
e23277. https://doi.org/10.1371/journal.pone.0023277.

Ismael, J. T. *How Physics Makes Us Free*. New York: Oxford University Press, 2016.

Ito, Masao. *The Cerebellum: Brain for an Implicit Self*. Upper Saddle River, New Jersey: FT Press, 2012.

James, William. *The Principles of Psychology: In Two Volumes*. Vol. 1. 1890. Facsim. ed. New York: Dover, 1995.

Jarvis, Erich D. "Evolution of Vocal Learning and Spoken Language." *Science* 366, no. 6461 (October 4, 2019): 50–54. https://doi.org /10.1126/science.aax0287.

Jaynes, Julian. *The Origin of Consciousness in the Breakdown of the Bicameral Mind*. New York: Houghton Mifflin, 2000.

Johnson, Eric. *Anxiety and the Equation: Understanding Boltzmann's Entropy*. Cambridge: MIT Press, 2018.

Kandel, Eric R., James H. Schwartz, Thomas M. Jessell, Steven A. Siegelbaum, and A. J. Hudspeth, eds. *Principles of Neural Science*. 5th edition. New York: McGraw-Hill, 2012.

Kant, Immanuel. *Critique of Pure Reason*. Edited by Marcus Weigelt. Translated by Friedrich Max Müller. Penguin Classics. London: Penguin Books, 2007.

Karinthy, Frigyes, and Oliver Sacks. *A Journey Round My Skull*. Translated by Vernon Duckworth Barker. New York: NYRB Classics, 2008.

Khatchadourian, Raffi. "How to Control a Machine with Your Brain." *New Yorker*, November 19, 2018. https://www .newyorker.com/magazine/2018/11/26/how-to-control-a -machine-with-your-brain.

Kirchhoff, Michael, Thomas Parr, Ensor Palacios, Karl Friston, and Julian Kiverstein. "The Markov Blankets of Life: Autonomy, Active Inference and the Free Energy Principle." *Journal of the Royal Society Interface* 15, no. 138 (January 2018): 20170792. https://doi.org/10.1098/rsif.2017.0792.

Kishita, Yuki, Hiroshi Ueda, and Makio Kashino. "Eye and Head Movements of Elite Baseball Players in Real Batting." *Frontiers in Sports and Active Living*, January 29, 2020. https://doi.org/10 .3389/fspor.2020.00003.

Knoll, Andrew H. *A Brief History of Earth: Four Billion Years in Eight Chapters*. New York: Custom House, 2021.

Koch, Christof. *Consciousness: Confessions of a Romantic Reduction-ist*. Cambridge: MIT Press, 2012.

———. *The Feeling of Life Itself: Why Consciousness Is Widespread but Can't Be Computed*. Ill. ed. Cambridge: MIT Press, 2019.

———. "Hot or Not." *Nature Human Behaviour* 4, no. 10 (October 2020): 991–92. https://doi.org/10.1038/s41562-020-0925-7.

Koch, Christof, and Klaus Hepp. "Quantum Mechanics in the Brain." *Nature* 440, no. 7084 (March 2006): 611. https://doi.org /10.1038/440611a.

Koch, Christof, and Patrick House. "Brain Bridging." *Nature*, August 26, 2020, d41586-020-02469-0. https://doi.org/10.1038/d41586 -020-02469-0.

Kristan, William B. "Early Evolution of Neurons." *Current Biology* 26, no. 20 (October 2016): R949–54. https://doi.org/10.1016/j .cub.2016.05.030.

Kuhn, Thomas S., and Ian Hacking. *The Structure of Scientific Revolutions*. 4th ed. Chicago: University of Chicago Press, 2012.

LaBerge, Stephen, Benjamin Baird, and Philip G. Zimbardo. "Smooth Tracking of Visual Targets Distinguishes Lucid REM Sleep Dreaming and Waking Perception from Imagination." *Nature Communications* 9, no. 1 (August 17, 2018): 3298. https://doi.org/10.1038/s41467-018-05547-0.

Land, M. F., and D. N. Lee. "Where We Look When We Steer." *Nature* 369, no. 6483 (June 30, 1994): 742–44. https://doi.org /10.1038/369742a0.

Land, M., N. Mennie, and J. Rusted. "The Roles of Vision and Eye Movements in the Control of Activities of Daily Living." *Perception* 28, no. 11 (1999): 1311–28. https://doi.org/10.1068/p2935.

Land, Michael F., and Dan-Eric Nilsson. *Animal Eyes*. 2nd ed. Oxford Animal Biology Series. Oxford: Oxford University Press, 2012.

Lane, Nick. *Life Ascending: The Ten Great Inventions of Evolution*. Ill. ed. New York: Norton, 2010.

———. *Oxygen: The Molecule That Made the World*. Rev. impression. Oxford Landmark Science Series. Oxford: Oxford University Press, 2016.

———. *Power, Sex, Suicide: Mitochondria and the Meaning of Life*. 2nd ed. Oxford Landmark Science Series. Oxford: Oxford University Press, 2018.

———. *The Vital Question: Energy, Evolution, and the Origins of Complex Life*. New York: Norton, 2016.

Larson, Erik J. *The Myth of Artificial Intelligence: Why Computers Can't Think the Way We Do*. Cambridge: Belknap Press of Harvard University Press, 2021.

LeDoux, Joseph E., and Caio Sorrentino. *The Deep History of Ourselves: The Four-Billion-Year Story of How We Got Conscious Brains*. New York: Viking, 2019.

Lee, Adam, Alison Nolan, Jason Watson, and Michael Tristem. "Identification of an Ancient Endogenous Retrovirus, Predating the Divergence of the Placental Mammals." *Philosophical Transactions of the Royal Society B: Biological Sciences* 368, no. 1626 (September 19, 2013): 20120503. https://doi.org/10.1098/rstb.2012.0503.

Leeuwen, Peter M. van, Stefan de Groot, Riender Happee, and Joost C. F. de Winter. "Differences between Racing and Non-Racing Drivers: A Simulator Study Using Eye-Tracking." *PLoS One* 12, no. 11 (November 9, 2017): e0186871. https://doi.org/10.1371/journal.pone.0186871.

Lemaitre, Anne-Laure, Marion Luyat, and Gilles Lafargue. "Individuals with Pronounced Schizotypal Traits Are Particularly Successful in Tickling Themselves." *Consciousness and Cognition* 41 (April 1, 2016): 64–71. https://doi.org/10.1016/j.concog.2016.02.005.

Levenson, Thomas. *The Hunt for Vulcan: . . . And How Albert Einstein Destroyed a Planet, Discovered Relativity, and Deciphered the Universe*. New York: Random House, 2016.

Lindley, David, and Ludwig Boltzmann. *Boltzmann's Atom: The Great Debate That Launched a Revolution in Physics*. New York: Free Press, 2001.

Lindsay, Grace. "Episode 25: What Can Eye Movements Tell Us About the Mind?" October 1, 2017. In *Unsupervised Thinking*. Podcast. http://unsupervisedthinkingpodcast.blogspot.com/2017/10/episode-25-what-can-eye-movements-tell.html.

Llinás, R., K. Walton, D. E. Hillman, and C. Sotelo. "Inferior Olive: Its Role in Motor Learning." *Science* 190, no. 4220 (December 19, 1975): 1230–31. https://doi.org/10.1126/science.128123.

Llinás, R., and J. P. Welsh. "On the Cerebellum and Motor Learning." *Current Opinion in Neurobiology* 3, no. 6 (December 1993): 958–65. https://doi.org/10.1016/0959-4388(93)90168-x.

Llinás, Rodolfo R. *I of the Vortex: From Neurons to Self*. Repr. ed. Cambridge, Mass.: Bradford Books, 2002.

———. "Intrinsic Electrical Properties of Mammalian Neurons and CNS Function: A Historical Perspective." *Frontiers in Cellular Neuroscience* 8 (2014): 320. https://doi.org/10.3389/fncel.2014.00320.

Llinás, Rodolfo, and Constantino Sotelo. *The Cerebellum Revisited*. New York: Springer US, 1992.

Luthra, Mahi, Eduardo J. Izquierdo, and Peter M. Todd. "Cognition Evolves with the Emergence of Environmental Patchiness." In *The 2020 Conference on Artificial Life*, 450–58. Online. MIT Press, 2020. https://doi.org/10.1162/isal_a_00330.

Maile, Amy E., Christopher G. Inoue, Larry E. Barksdale, and David O. Carter. "Toward a Universal Equation to Estimate Postmortem Interval." *Forensic Science International* 272 (March 2017): 150–53. https://doi.org/10.1016/j.forsciint.2017.01.013.

Maletic, Greg, Duncan Brown, Larry DeMar, and George Gomez. *The Future of Pinball*. Documentary, n.d.

Martinez, Pedro, and Simon G. Sprecher. "Of Circuits and Brains: The Origin and Diversification of Neural Architectures." *Frontiers in Ecology and Evolution* 8 (2020): 82. https://doi.org/10.3389/fevo.2020.00082.

Mascetti, Gian Gastone. "Unihemispheric Sleep and Asymmetrical Sleep: Behavioral, Neurophysiological, and Functional Perspectives." *Nature and Science of Sleep* 8 (July 12, 2016): 221–38. https://doi.org/10.2147/NSS.S71970.

Mashour, George Alexander, and Ralph Lydic. *Neuroscientific Foundations of Anesthesiology*. New York: Oxford University Press, 2011.

Massimini, Marcello, and Giulio Tononi. *Sizing Up Consciousness: Towards an Objective Measure of the Capacity for Experience*. Ill. ed. Oxford: Oxford University Press, 2018.

McNamara, Patrick. *The Neuroscience of Sleep and Dreams*. Cambridge Fundamentals of Neuroscience in Psychology. Cambridge: Cambridge University Press, 2019.

McPhee, John. "The Pinball Philosophy." *New Yorker*, June 22, 1975. http://www.newyorker.com/magazine/1975/06/30/the-pinball-philosophy.

Mildner, Judith N., and Diana I. Tamir. "Spontaneous Thought as an Unconstrained Memory Process." *Trends in Neurosciences* 42, no. 11 (November 2019): 763–77. https://doi.org/10.1016/j.tins.2019.09.001.

Mochizuki, Yasuhiro, and Shigeru Shinomoto. "Analog and Digital Codes in the Brain." *Physical Review E* 89, no. 2 (February 4, 2014): 022705. https://doi.org/10.1103/PhysRevE.89.022705.

Morgan, Charity J., Michael J. Coleman, Ayse Ulgen, Lenore Boling,

Jonathan O. Cole, Frederick V. Johnson, Jan Lerbinger, J. Alexander Bodkin, Philip S. Holzman, and Deborah L. Levy. "Thought Disorder in Schizophrenia and Bipolar Disorder Probands, Their Relatives, and Nonpsychiatric Controls." *Schizophrenia Bulletin* 43, no. 3 (May 2017): 523–35. https://doi.org/10.1093/schbul /sbx016.

Moser, Edvard I., May-Britt Moser, and Yasser Roudi. "Network Mechanisms of Grid Cells." *Philosophical Transactions of the Royal Society B: Biological Sciences* 369, no. 1635 (February 5, 2014): 20120511. https://doi.org/10.1098/rstb.2012.0511.

Moser, May-Britt, and Edvard I. Moser. "Crystals of the Brain." *EMBO Molecular Medicine* 3, no. 2 (February 2011): 69–71. https://doi.org/10.1002/emmm.201000118.

Moser, May-Britt, David C. Rowland, and Edvard I. Moser. "Place Cells, Grid Cells, and Memory." *Cold Spring Harbor Perspectives in Biology* 7, no. 2 (February 2, 2015): a021808. https://doi.org /10.1101/cshperspect.a021808.

Mota-Rolim, Sergio Arthuro. "On Moving the Eyes to Flag Lucid Dreaming." *Frontiers in Neuroscience* 14 (April 15, 2020): 361. https://doi.org/10.3389/fnins.2020.00361.

Murphey, R. K., and John Palka. "Efferent Control of Cricket Giant Fibres." *Nature* 248, no. 5445 (March 1974): 249–51. https://doi .org/10.1038/248249a0.

Musha, T., and M. Yamamoto. "1/f Fluctuations in Biological Systems." In *Proceedings of the 19th Annual International Conference of the IEEE Engineering in Medicine and Biology Society: "Magnificent Milestones and Emerging Opportunities in Medical Engineering"* 6: 2692–97, 1997. https://doi.org/10.1109/IEMBS.1997.756890.

Neal, Meghan. "The Many Colors of Sound." *Atlantic*, February 16, 2016. https://www.theatlantic.com/science/archive/2016/02 /white-noise-sound-colors/462972/.

Nocke, Harald. "Physiological Aspects of Sound Communication in Crickets (Gryllus Campestris L.)." *Journal of Comparative Physiology* 80, no. 2 (June 1, 1972): 141–62. https://doi.org/10 .1007/BF00696487.

Nokes, L. D., T. Flint, J. H. Williams, and B. H. Knight. "The Application of Eight Reported Temperature-Based Algorithms to Calculate the Postmortem Interval." *Forensic Science International* 54, no. 2 (May 1992): 109–25. https://doi.org/10.1016 /0379–0738(92)90155-p.

"Off Road, but Not Offline: How Simulation Helps Advance Our Waymo Driver." Waymo. April 28, 2020. https://blog.waymo.com /2020/04/off-road-but-not-offline--simulation27.html.

Ohno, Susumu. *Evolution by Gene Duplication*. London: Allen and Unwin; Springer Verlag, 1970.

Ólafsdóttir, H. Freyja, Daniel Bush, and Caswell Barry. "The Role of Hippocampal Replay in Memory and Planning." *Current Biology* 28, no. 1 (January 8, 2018): R37–50. https://doi.org/10.1016/j .cub.2017.10.073.

"OpenAI Five." June 25, 2018. https://openai.com/blog/openai-five/.

O'Regan, J. Kevin. *Why Red Doesn't Sound Like a Bell: Understanding the Feel of Consciousness*. Ill. ed. New York: Oxford University Press, 2011.

Ortiz de Gortari, Angelica B., and Mark D. Griffiths. "Game Transfer Phenomena and Its Associated Factors: An Exploratory Empirical Online Survey Study." *Computers in Human Behavior* 51 (October 1, 2015): 195–202. https://doi.org/10.1016/j.chb.2015 .04.060.

Pang, Rich, Benjamin J. Lansdell, and Adrienne L. Fairhall. "Dimensionality Reduction in Neuroscience." *Current Biology* 26, no. 14 (July 25, 2016): R656–60. https://doi.org/10.1016/j.cub.2016 .05.029.

Parent, Andre, and Malcolm B. Carpenter. *Carpenter's Human Neuroanatomy*. Subsequent ed. Baltimore: Williams and Wilkins, 1996.

Pastuzyn, Elissa D., et al. "The Neuronal Gene Arc Encodes a Repurposed Retrotransposon Gag Protein That Mediates Intercellular RNA Transfer." *Cell* 172, no. 1 (January 11, 2018): 275–288. e18. https://doi.org/10.1016/j.cell.2017.12.024.

Pellicer, Jaume, Michael F. Fay, and Ilia J. Leitch. "The Largest Eukaryotic Genome of Them All?" *Botanical Journal of the Linnean Society* 164, no. 1 (September 1, 2010): 10–15. https://doi.org /10.1111/j.1095–8339.2010.01072.x.

Penrose, Roger. *Shadows of the Mind: A Search for the Missing Science of Consciousness*. Repr. ed. Oxford: Oxford University Press, 1996.

Petronius Arbiter, and William Arrowsmith. *The Satyricon*. New York: Meridian, 1994.

Pinsky, Robert. *The Sounds of Poetry: A Brief Guide*. New York: Farrar, Straus and Giroux, 1999.

Plato. *Republic*.

Prescott, Tony J., Nathan Lepora, and Paul F. M. J. Verschure, eds. *Living Machines: A Handbook of Research in Biomimetics and Biohybrid Systems*. New York: Oxford University Press, 2018.

Prieur, Jacques, Stéphanie Barbu, Catherine Blois-Heulin, and Alban Lemasson. "The Origins of Gestures and Language: History, Current Advances and Proposed Theories." *Biological Reviews* 95, no. 3 (2020): 531–54. https://doi.org/10.1111/brv.12576.

Quammen, David. *The Tangled Tree: A Radical New History of Life*. New York: Simon & Schuster, 2018.

Raffone, Antonino, Narayanan Srinivasan, and Cees van Leeuwen. "Perceptual Awareness and Its Neural Basis: Bridging Experimental and Theoretical Paradigms." *Philosophical Transactions of the Royal Society B: Biological Sciences* 369, no. 1641 (May 5, 2014): 20130203. https://doi.org/10.1098/rstb.2013.0203.

"Rare Japanese Plant Has Largest Genome Known to Science." ScienceDaily. Accessed October 10, 2021. https://www.sciencedaily.com/releases/2010/10/101007120641.htm.

Reardon, Sara. "Cells Hack Virus-like Protein to Communicate." *Nature*, January 11, 2018. https://doi.org/10.1038/d41586-018-00492-w.

Roitblat, H. L, and Jean-Arcady Meyer. *Comparative Approaches to Cognitive Science*. Cambridge: MIT Press, 1995.

Sacks, Oliver. *Awakenings*. New York: Vintage Books, 1999.

Sample, Ian. "'Eureka Machine' Puts Scientists in the Shade by Working Out Laws of Nature." *Guardian*, April 3, 2009. https://www.theguardian.com/science/2009/apr/02/eureka-laws-nature-artificial-intelligence-ai.

Sapolsky, Robert M. *Behave: The Biology of Humans at Our Best and Worst*. New York: Penguin Press, 2017.

Scholes, Chris, Paul V. McGraw, and Neil W. Roach. "Learning to Silence Saccadic Suppression." *Proceedings of the National Academy of Sciences* 118, no. 6 (February 9, 2021). https://doi.org/10.1073/pnas.2012937118.

Schrödinger, Erwin. "*What Is Life? The Physical Aspect of the Living Cell*" with "*Mind and Matter*" and "*Autobiographical Sketches*." Canto Classics. Cambridge: Cambridge University Press, 1992.

Schwarzlose, Rebecca. *Brainscapes: The Warped, Wondrous Maps Written in Your Brain—And How They Guide You*. Boston: Mariner Books, 2021.

Selimbeyoglu, Aslihan, and Josef Parvizi. "Electrical Stimulation of the Human Brain: Perceptual and Behavioral Phenomena Reported in the Old and New Literature." *Frontiers in Human Neuroscience* 4 (2010): 46. https://doi.org/10.3389/fnhum.2010.00046.

Shalhoub, Michael. *The Pinball Compendium: Electro-Mechanical Era*. Ill. ed. Atglen, Pa.: Schiffer Publishing, 2008.

Shank, Matthew D., and Kathleen M. Haywood. "Eye Movements While Viewing a Baseball Pitch." *Perceptual and Motor Skills* 64, no. 3 suppl. (June 1, 1987): 1191–97. https://doi.org/10.2466 /pms.1987.64.3c.1191.

"Shocked Moth Remembers Past Life as Caterpillar!" Accessed October 8, 2021. https://www.science.org/content/article/shocked -moth-remembers-past-life-caterpillar.

Shrestha, Rijen, Tanuj Kanchan, and Kewal Krishan. "Methods of Estimation of Time Since Death." In *StatPearls*. Treasure Island, Fla.: StatPearls Publishing, 2021. http://www.ncbi.nlm.nih.gov /books/NBK549867/.

Smith, Daniel B. *Muses, Madmen, and Prophets: Hearing Voices and the Borders of Sanity*. New York: Penguin, 2008.

Smith, Stephen J., Michael Hawrylycz, Jean Rossier, and Uygar Sümbül. "New Light on Cortical Neuropeptides and Synaptic Network Plasticity." *Current Opinion in Neurobiology* 63 (August 1, 2020): 176–88.

Sober, Elliott. *Ockham's Razors: A User's Manual*. Cambridge: Cambridge University Press, 2015.

Springer, Mark S. et al. "Genomic and Anatomical Comparisons of Skin Support Independent Adaptation to Life in Water by Cetaceans and Hippos." *Current Biology* 31, no. 10 (May 24, 2021): 2124–2139.e3. https://doi.org/10.1016/j.cub.2021.02.057.

Stensola, Hanne, Tor Stensola, Trygve Solstad, Kristian Frøland, May-Britt Moser, and Edvard I. Moser. "The Entorhinal Grid Map Is Discretized." *Nature* 492, no. 7427 (December 2012): 72–78. https://doi.org/10.1038/nature11649.

Stevens, Craig W. "The Evolution of Vertebrate Opioid Receptors." *Frontiers in Bioscience: A Journal and Virtual Library* 14 (January 1, 2009): 1247–69.

Stickgold, Robert, April Malia, Denise Maguire, David Roddenberry, and Margaret O'Connor. "Replaying the Game: Hypnagogic Images in Normals and Amnesics." *Science* 290, no. 5490 (October 13, 2000): 350–53. https://doi.org/10.1126/science.290.5490.350.

Stirman, Shannon, and James Pennebaker. "Word Use in the Poetry of Suicidal and Nonsuicidal Poets." *Psychosomatic Medicine* 63 (July 1, 2001): 517–22. https://doi.org/10.1097/00006842-200107000-00001.

Stoppard, Tom. *The Hard Problem*. New York: Grove Press, 2015.

Straka, Hans, John Simmers, and Boris P. Chagnaud. "A New Perspective on Predictive Motor Signaling." *Current Biology* 28, no. 5 (March 2018): R232–43. https://doi.org/10.1016/j.cub.2018.01.033.

Suddendorf, Thomas, Donna Rose Addis, and Michael C. Corballis. "Mental Time Travel and the Shaping of the Human Mind." *Philosophical Transactions of the Royal Society B: Biological Sciences* 364, no. 1521 (May 12, 2009): 1317–24. https://doi.org/10.1098/rstb.2008.0301.

Suddendorf, Thomas, and Michael C. Corballis. "The Evolution of Foresight: What Is Mental Time Travel, and Is It Unique to Humans?" *Behavioral and Brain Sciences* 30, no. 3 (June 2007): 299–313. https://doi.org/10.1017/S0140525X07001975.

Tegmark, Max. "Consciousness as a State of Matter." *Chaos, Solitons and Fractals* 76 (July 2015): 238–70. https://doi.org/10.1016/j.chaos.2015.03.014.

Tiessen, Paul. "A Comparative Approach to the Form and Function of Novel and Film: Dorothy Richardson's Theory of Art." *Literature/Film Quarterly* 3, no. 1 (1975): 83–90.

Tong, Frank, Ming Meng, and Randolph Blake. "Neural Bases of Binocular Rivalry." *Trends in Cognitive Sciences* 10, no. 11 (November 2006): 502–11. https://doi.org/10.1016/j.tics.2006.09.003.

Tononi, Giulio. *Phi: A Voyage from the Brain to the Soul*. New York: Pantheon, 2012.

Tononi, Giulio, and Gerald M. Edelman. "Consciousness and Complexity." *Science* 282, no. 5395 (December 4, 1998): 1846–51. https://doi.org/10.1126/science.282.5395.1846.

Tononi, Giulio, and Christof Koch. "Consciousness: Here, There and Everywhere?" *Philosophical Transactions of the Royal Society B: Biological Sciences* 370, no. 1668 (May 19, 2015): 20140167. https://doi.org/10.1098/rstb.2014.0167.

Van den Pol, Anthony N. "Neuropeptide Transmission in Brain Circuits." *Neuron* 76, no. 1 (October 4, 2012): 98–115. https://doi.org/10.1016/j.neuron.2012.09.014.

Vuilleumier, Patrik. "Anosognosia: The Neurology of Beliefs and Uncertainties." *Cortex* 40, no. 1 (January 1, 2004): 9–17. https://doi.org/10.1016/S0010–9452(08)70918-3.

Wang, Jane, Zeb Kurth-Nelson, and Matt Botvinick, "Prefrontal Cortex as a Meta-reinforcement Learning System." *Deepmind*. May 14, 2018. https://deepmind.com/blog/article/prefrontal-cortex-meta-reinforcement-learning-system.

Weinberger, Eliot. *Nineteen Ways of Looking at Wang Wei*. Repr. ed. New York: New Directions, 2016.

Whyte, Christopher J., and Ryan Smith. "The Predictive Global Neuronal Workspace: A Formal Active Inference Model of Visual Consciousness." February 12, 2020. https://doi.org/10.1101/2020.02.11.944611.

Wissner-Gross, Alex. "A New Equation for Intelligence." TED Talk, 2014. https://www.youtube.com/watch?v=ue2ZEmTJ_Xo.

Wittgenstein, Ludwig. *Tractatus Logico-Philosophicus*. 2nd ed. London: Routledge, 2001.

Witthoft, Nathan, and Jonathan Winawer. "Learning, Memory, and Synesthesia." *Psychological Science* 24, no. 3 (March 1, 2013): 258–65. https://doi.org/10.1177/0956797612452573.

Wolman, David. "The Split Brain: A Tale of Two Halves." *Nature* 483, no. 7389 (March 1, 2012): 260–63. https://doi.org/10.1038/483260a.

Index

· · · · · ·

Tyler MacNiven

PATRICK HOUSE is a neuroscientist with a Ph.D. from Stanford University. His scientific research focuses on the neuroscience of free will and how mind-control parasites alter their host's behavior. He writes about science, technology, and culture for newyorker.com, *Nautilus* magazine, and *Slate*; his work was selected for the *Best American Science and Nature Writing 2020* anthology. His fiction has been published in *Nature*. House lives in Los Angeles, California.